Astrophysics

by Cynthia Phillips, PhD
and Shana Priwer

A Wiley Brand

Astrophysics For Dummies®

Published by: **John Wiley & Sons, Inc.**, 111 River Street, Hoboken, NJ 07030-5774, www.wiley.com

Copyright © 2024 by John Wiley & Sons, Inc., Hoboken, New Jersey

Published simultaneously in Canada

For general information on our other products and services, please contact our Customer Care Department within the U.S. at 877-762-2974, outside the U.S. at 317-572-3993, or fax 317-572-4002. For technical support, please visit https://hub.wiley.com/community/support/dummies.

Wiley publishes in a variety of print and electronic formats and by print-on-demand. Some material included with standard print versions of this book may not be included in e-books or in print-on-demand. If this book refers to media such as a CD or DVD that is not included in the version you purchased, you may download this material at http://booksupport.wiley.com. For more information about Wiley products, visit www.wiley.com.

Library of Congress Control Number: 2024931757

ISBN 978-1-394-23504-9 (pbk); ISBN 978-1-394-23505-6 (ebk); ISBN 978-1-394-23506-3 (ebk)

SKY10067066_021424

Contents at a Glance

Table of Contents

Introduction

I f you spend your evenings pondering the constellations and spotting shooting stars, you're already hooked on astronomy.

If you're one of those people who "gets" how things work, and if you've ever mentioned Newton or Einstein in casual conversation, apologize to your friends because you've got a keen interest in physics and are willing to share it.

Put the two together, and welcome to astrophysics! This field combines the excitement of recognizing planets and stars with the satisfaction of applying mathematics and physics to those same objects. Astrophysics goes beyond cataloging and observing the night sky into performing calculations, making measurements and creating predictions about future behaviors.

As a field, astrophysics covers everything you can see in the sky and then some. From the smallest molecule (no, you can't see these without specialized equipment) to stars and planets, individual celestial bodies have a creation story that mimics that of the universe's evolution. And the cosmos certainly isn't limited to planets and stars!

For example, nebulae are massive dust and gas clouds that often result from the explosive death of a star but, coincidentally (actually not a coincidence, as you'll read in Chapter 5), they're also regions where new stars are born. Galaxies are vast groupings of stars, often containing nebulae and other high-energy astronomical features such as black holes and neutron stars, and they group together to form clusters of galaxies. Nothing in the universe exists in isolation, not even quarks (these are the smallest particles in the universe. Curious? Check out Chapter 15).

And speaking of things that are infinitely tiny, the Big Bang is the both the ending and the starting point of our exploration into astrophysics. The universe began as a single point in space and time, growing explosively over subsequent seconds as it expanded. The forces at play in our current universe (think gravity, electromagnetism, and the nuclear forces) emerged, atoms began to form, matter separated from radiation, and every element in today's world was created. From the cloud of cosmic dust emerged stars, galaxies, dark matter, and all the other pieces of the cosmic puzzle.

But eventually, all good things must come to an end, and unfortunately that includes our universe. Theories abound as to how we'll all meet our demise, but those theories only open the door to more questions. Is this universe the only one to have ever existed? Will another one be created when this one's gone? Or, as some theories suggest, is our universe only one in a series? There are no definitive answers at this point to many of these questions, but studying astrophysics gives you the knowledge you need to start asking those questions, and perhaps even answering those questions for yourself.

About This Book

Welcome to the cosmos! Whether you're curious about how the universe began, want to know more about the science behind eclipses, or are considering becoming an astrophysicist (and if that's the case, we expect to see you at conferences and lectures down the road), this book will get you started.

Astrophysics is a notoriously difficult idea to wrap your head around because it's incredibly vast, yet also extremely detail-oriented. We've broken down the information into digestible chapters so that you can read this book all at once, or can flip through to only the sections that interest you. We've written each chapter as a stand-alone piece with references to other parts of the book as needed. If you have the time and interest, try reading the book cover to cover for a more complete sense of the story: Decipher what you observe in the night sky, understand the science behind it, and get a holistic overview of this universe we call home, from start to finish.

Got questions on unfamiliar terms? Check the glossary at the back of this book for a quick guide. Need a quick refresher on which formula goes with which concept?

Finally, please consider us as resources for any additional questions you may have. Wiley can put you in contact with us directly for follow-up questions, and you can always request additional books and content directly from the publisher. We want to share our love of space science with you! Let us know how we can help.

Foolish Assumptions

We assume you have picked up this book because you think astrophysics sounds interesting — dare we say cool? Maybe you are concerned about the end of the universe. Maybe you want to know what you're looking for at a star party, or perhaps you're looking for a bit of light bedtime reading (hint: maybe save the end-of-the-world chapters for the morning). Any reason to learn about astrophysics is a good one, and we hope this book has fun and thought-provoking information for you.

We don't assume that you are a scientist or have even taken any science classes since high school — or maybe you're a high school student now! Whatever your starting point, this book is for you.

A few key tips before we get started. Scientists, no matter their location, typically use the metric system when running physics or astronomy calculations. The metric system is also known as the International System of units, or SI units. Although you may be more accustomed to the standard American system of feet and pounds (amusingly, this system is actually called the Imperial system!), in this book we provide key quantities in both systems of measurement. Don't worry, though, we won't make you convert the speed of light into furlongs per fortnight! (If a conversion like that is needed, we provide it for you.)

Many of the key words used in this book derive from languages more ancient than modern-day English. *Nebula* for example, comes from Latin and its plural in Latin is *nebulae*. We use the language-appropriate plural for these words throughout the book rather than improper Englishizations (you'll not see reference to *nebulas*). Similarly, scientists prefer standard units of time, and so we refer to CE (common era, starting at the year zero) and BCE (before common era, starting before the year zero).

Icons Used in This Book

Throughout this book, helpful icons can guide you to particularly useful nuggets of wisdom, and also help you see what's fine to skip if your eyes are glazing over. Here's what each symbol means:

REMEMBER

The string-tied-on-a-finger icon points out information that will be useful to remember for the future.

This icon indicates technical info; the content next to these icons will typically be information related to the topic you're reading about but with a more in-depth technical explanation. Feel free to skip over these if you're looking for the bigger picture, but use these callouts when you're searching for more detailed information on a subject.

The light bulb graphic highlights particularly useful or interesting tidbits. Scan the page quickly and let your eye be drawn to the Tips for nuggets of information — such as, for example, how stars created light in the first place!

From black holes to solar eclipses to getting tangled up in the math of general relativity, astrophysics can head into complex and dangerous territory. The warning icon calls out areas that may be dangerous (either intellectually or physically!) and require a careful approach.

Beyond the Book

In addition to the book you're reading right now, be sure to check out the free Cheat Sheet. It offers a timeline of astrophysics discoveries, a list of misconceptions, and a list of world record holders, among other things. To get this Cheat Sheet, simply go to www.dummies.com and enter **Astrophysics For Dummies** in the Search box.

There are also some other books in this series that you might enjoy. *Astronomy For Dummies*, by Stephen Maran and Richard Fienberg, could be a great starting point if you want to learn more about the observational side of astronomy. You can also brush up on your physics knowledge in *Physics I For Dummies* or *Physics II For Dummies* by Steven Holzner.

Where to Go from Here

We're overjoyed to greet you on your journey into astrophysics. If this book were a welcome into our home, you'd be greeted with the aroma of chocolate chip cookies fresh from the oven. We suppose the information contained inside may be a poor substitute, but we've done our best to provide an overview of astrophysics using the written word instead of chocolate.

Feel free to consume this book all at once or sample as you go. Looking for a basic overview of astronomy and physics before you get in too deep? Try out Chapter 2 and Chapter 3 for a quick refresher on everything you wanted to know about astronomy and physics but might have forgotten or been afraid to ask.

How did those stars form? Chapter 5 has you covered. Got a yearning for black holes? Don't worry, Chapter 8 is ready to suck you in (literally? You be the judge). How did it all begin, and how will it end? The Big Bang starts in Chapter 13, then concludes in Chapter 16 with the end of it all.

The universe seems to have had a beginning, a middle (we're in it now), and an end. As you'll learn between these pages, there's a lot going on in between. We hope that by the time you reach the end of your exploration, you'll be able to look up at those stars with a sense of both awe and understanding. Welcome to astrophysics!

1

Getting Started with Astrophysics

Refresh your understanding of matter, forces, and energy — the fundamental laws of physics.

Discover the world of astronomy, from our solar system to our galaxy, and see how the constellations fit in.

Gain a big-picture understanding of how astrophysics bridges the gap between astronomy and physics. Learn how telescopes are used, and study the Sun as a star to find out more about space weather and eclipses.

Chapter 1

Welcome to the Universe

Have you ever looked up at the night sky and wondered what you're seeing? How did all those dots in the sky appear? Why are some brighter than others? Have you felt that sense of awe and wonder deep in your soul, and realized that you're only a small part of something much greater? If any of these apply to you, welcome to *Astrophysics For Dummies!*

You're in good company with your pondering of the universe. Since our earliest surviving records, humans have shared this fascination with the cosmos. And, fortunately, information and knowledge about the universe are exponentially greater today than they were during the time of our ancestors.

Although simply gazing up at the heavens can be inspiring, understanding what you are looking at can make the experience mind-blowing. Gazing at the sky reveals not only other worlds in our solar system but also other stars, many of which may have planets of their own. For example, if the sky is dark enough you can see the Milky Way, the bright band of stars stretching across the sky that's actually the disk of the Milky Way Galaxy. Your knowledge of astrophysics turns a beautiful spectacle into something known but no less amazing.

TIP

With good eyesight or some binoculars and/or a telescope, you can start to see nebulae, and understand that many are clouds of gas and dust that are stellar nurseries, where new stars are being born. You can even see galaxies beyond the Milky Way, and realize that they contain billions of their own stars. Due to a fundamental property of astrophysics (the speed of light being a constant), the vast cosmic distances to these objects also means that when you see them, you are actually looking back in time. Astrophysics can then be seen as a study of time as well as of space, and it can take you all the way back to the dawn of time itself — the Big Bang, the event that created our universe.

The word *astrophysics* may seem daunting, but it's nothing more than a scientific term combining a descriptive view of the universe (that's the astronomy part) with a mathematical understanding of the theoretical basis for what you are seeing (that's the physics part). Don't worry; we bring you up to speed on the fundamentals before diving into the details of the universe. Before you know it, you'll be able to explain and understand your place in the cosmos, and know a bit more about how the world works.

Welcome to astrophysics!

The Science of Astrophysics

It's only been within the last 150 years that the field of astrophysics has really taken off as separate from either astronomy or physics. Astronomy is essentially a science of observation, whereas astrophysics is more concerned with understanding those observations. Gear up and let's dive in!

The start of astronomy

TIP

The first few thousand years of astronomy can be seen as largely descriptive. Humans around the world documented the sky, observed changes, and made up stories to explain what they saw. These stories were recorded into the names of the constellations, and they were created by cultures all around the world. People observed that although most stars had fixed patterns in the sky, there were also repeating patterns. The Sun rose and set predictably, for example. Early observers noted a few interlopers: Stars that changed position over the course of the year were later shown to be planets, and flashy visitors such as the occasional comet and meteor made their own appearances.

As time went on, astronomical observations became more rigorous as telescopes were invented and used to observe the sky in more detail. Astronomers soon discovered that there was more to the sky than sparking points of light. Although most of these objects were stars, 19th-century telescopes and the discovery of photography revealed the larger, fainter, and fuzzier objects as nebulae and galaxies. With this expanded cast, the stage was set, and interest in the cosmos was sufficiently piqued to incite an entirely new field of study, one that stretched both imaginations and creativity to the maximum.

A beautiful connection: Physics, astronomy, and astrophysics

Physics, as you might recall from high school, is the study of how the natural world works. If you drop a can of beans on your toe, that's gravity at work. Astronomy, on the other hand, is the study of everything in the sky, from planets to stars to galaxies. Astrophysics joins the party as a more quantitative study that combines the observations of astronomy ("what") with the underlying theories of physics ("how"). Put simply, astrophysics is the study of how the cosmos works from beginning to end.

TIP

Astrophysics is, in many ways, a field that focuses on studying the intangible. Astrophysicists need to come up with specific ways of gathering information about phenomena that are, quite literally, out of this world. There are several ways in which scientists can tackle this problem:

>> **Observations:** Using Earth- and space-based telescopes and instrumentation, astrophysicists observe the universe at different wavelengths.

>> **Laboratory work:** Specially-designed equipment allows astrophysicists to simulate certain aspects of the cosmos right here at home. Assuming your home is an advanced-science lab, of course.

>> **Theory:** More than chalk on a blackboard, state-of-the-art supercomputers are used to run simulations on everything from the birth of a star to the end of the universe.

Check out Chapter 4 for more information on each of these concepts.

Let there be light! The electromagnetic spectrum

The observational part of astrophysics requires — surprise! — observations.

Astrophysicists observe the universe using a variety of methods. Because we can't (yet!) travel to other stars and galaxies, these observations are all based on detectable information that distant objects send out into space. Most of this information comes in the form of electromagnetic radiation.

Electromagnetic radiation (commonly known as *light*) is a way that energy travels through space, and it's a critical concept for anyone conducting astrophysical observations. The world visible to humans comprises only a small portion of what scientists call the *electromagnetic spectrum*, a way of describing all types of electromagnetic radiation in the universe.

The electromagnetic spectrum consists of seven classes of electromagnetic waves (all defined in terms of meters):

>> **Gamma rays:** Shorter than 1×10^{-11} meters

>> **X-rays:** 1×10^{-11} meters to 1×10^{-8} meters

>> **Ultraviolet (UV) light:** 1×10^{-8} meters to 4×10^{-7} meters

>> **Visible light (optical):** 4×10^{-7} meters to 7×10^{-7} meters

>> **Infrared:** 7×10^{-7} meters to 1×10^{-3} meters

>> **Microwaves:** 1×10^{-3} meters to 1×10^{-1} meters

>> **Radio waves:** Longer than 1×10^{-1} meters

These types of radiation are sorted by wavelength. The shorter the wavelength, the higher the energy. Gamma rays are the highest-energy type of radiation but they also have the shortest wavelength. This sorting of electromagnetic radiation in order by wavelength, is what's called the *electromagnetic (EM) spectrum*; see Figure 1-1.

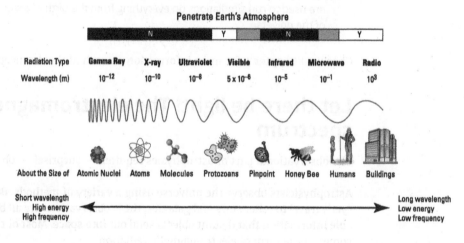

FIGURE 1-1:
The electromagnetic spectrum.

Electromagnetic radiation is carried by a particle called a *photon*, and the energy and wavelength of a photon are related by this simple equation:

$$E = \frac{hc}{\lambda}$$

In this equation, E is energy, h is a constant called Planck's constant, c is the speed of light, and λ (the Greek letter lambda) is the wavelength. You can see from this equation that energy is inversely proportional to wavelength, because wavelength is on the bottom of the fraction. As wavelength shrinks, energy grows.

TIP

And what is electromagnetic radiation? It's a way that photons, in the form of electromagnetic waves, travel through space. These waves carry both energy and momentum, and they can travel through the vacuum of space and through some materials. Visible light is a kind of electromagnetic radiation, as you can see in Figure 1-1, but so are radio waves, x-rays, and other kinds of familiar radiation.

Making waves

Wavelengths are what let us see colors. If you've ever looked up at the sky after a rainstorm to see a beautiful rainbow, that beautiful arch is caused by tiny droplets of water splitting visible light apart into all its different colors — the colors of the rainbow! The violet light you see has the shortest visible wavelength, green is in the middle, and red has the longest wavelength. As you'll learn in this book, the idea of colors can be extended broadly across the electromagnetic spectrum. The wavelengths of light given off or reflected by an object are related to its composition and can also be used to find its velocity and distance from us.

TECHNICAL
STUFF

This simple wavelength story might be starting to make sense but like many ideas in astrophysics, it's a bit more complicated than it seems. Light in particular, and electromagnetic radiation in general, have aspects of both a wave and a particle. Light can also travel through a vacuum at the speed of light (shocking, isn't it, to hear that the speed at which light travels is the speed of light?). As it turns out, the speed of light is a fundamental, fixed constant — nothing can go faster than light, and light (which is electromagnetic radiation after all!) always travels at this speed through a vacuum.

We refer to various parts of the electromagnetic spectrum throughout this book because different celestial bodies in space make their presence known in various ways. Stars, for example, emit mostly visible light that we can easily detect, but other objects such as neutron stars emit gamma rays.

TIP

Celestial bodies emit more than one type of radiation. Figure 1-2 shows a famous cloud of gas and dust called the Crab Nebula as viewed through telescopes at five different wavelengths, from the radio to the visible to the x-ray. See the color photo section for a beautiful multi-wavelength composite version.

FIGURE 1-2:
The Crab Nebula emitting radiation at different wavelengths.

Radio   Infrared   Optical   Ultraviolet   X-Ray

*Courtesy of G. Dubner (IAFE, CONICET-University of Buenos Aires) et al.; NRAO/AUI/NSF; A. Loll et al.; T. Temim et al.;
F. Seward et al.; Chandra/CXC; Spitzer/JPL-Caltech; XMM-Newton/ESA; and Hubble/STScI*

Astrophysicists use different kinds of telescopes, both on Earth and in space, to observe at wavelengths across the electromagnetic spectrum. It's often the combination of datasets taken at different wavelengths, such as in Figure 1-2, that yields new insights into how the cosmos operates.

TECHNICAL
STUFF

When you get to shorter wavelengths, astronomers sometimes use frequency instead to define them. Frequency is just defined as the number of wave cycles per second, so it is inversely related to wavelength. As the wavelength increases, the frequency decreases. In fact, for light and other kinds of electromagnetic radiation that travel at the fixed speed of light, the relationship can be expressed as

$$c = \lambda v$$

where c is the speed of light, λ is the wavelength, and v (the Greek letter nu) is the frequency. Wavelength is expressed in units of length (usually meters or nanometers), whereas frequency is expressed in units of Hertz (one Hertz means one cycle per second).

LIGHT VERSUS SOUND

It's not just light that comes in waves! Another highly familiar type of wave is the sound wave. Sound is a completely different process (and a completely different kind of wave) than electromagnetic radiation. The biggest difference is that sound requires a medium to move through in order to travel. Sound can be transmitted through air but also through the vibration of a musical instrument or even through the floor from your downstairs neighbor's speakers. Unlike light, however, sound can't travel through a vacuum. All those science fiction movies with ships whooshing through space are just that — fiction.

Tools of the Trade

No matter how good your eyesight may be, you'll never be able to see anything in deep space unaided. Unless you've got superpowers, you'll also have trouble seeing gamma rays, x-rays, or radio waves with the naked eye. Celestial objects such as pulsars and black hole accretion disks emit x-rays, for example, and x-rays are invisible to the human eye at the short end of the electromagnetic spectrum. There would be no realistic way to learn about these types of objects without specialized equipment. The following sections describe the most common types of observing tools astrophysicists require, and the differences between them.

The nitty-gritty of telescopes and astronomical instruments

If you've ever compared your view of the night sky from a major city to the countryside or desert, you know that the darker the sky, the more stars you can see. Astronomers take the "dark sky is better" concept to the next level when they are locating observations. Although your unaided eyes, binoculars, or a small telescope are great for a preliminary tour of the sky, performing the observational science that's key to astrophysics requires using a bigger telescope.

TIP

Is bigger better when it comes to telescopes? Absolutely, because most professional telescopes gather starlight using a mirror; larger-diameter mirrors gather more starlight that in turn allows you to see objects that are fainter or farther away.

Most professional optical astronomical observatories are located on the tops of mountains, as far away from civilizations as possible, for two reasons:

>> Mountains are usually some distance away from big cities. The skies are darker because there's less light pollution from city lights.

>> Mountaintops are typically at higher elevations than cities (unless you're someplace like Denver, Colorado, the "Mile High City.") The atmosphere is thinner the higher you go, creating less air and stuff like water vapor between you and the stars. Because the atmosphere is constantly in motion, more atmosphere can mean blurry images, and water vapor blocks some colors of light. The less atmosphere, the better.

Some types of observatories don't actually require a long winding mountain road for access. Radio telescopes can be located at sea level. They also need to be away from civilization, though, because they're extremely sensitive to interference in the radio portion of the EM spectrum. If a scientist is observing with an optical telescope, they have to be careful to keep flashlights away from the telescopes because that light would interfere with viewing.

At a radio telescope observatory, though, cell phones are banned because the radiation they give off interferes with those radio telescopes. Solar telescopes, on the other hand, operate during daytime and don't need a dark sky location; these telescopes often use special filters to dim our Sun's intense light enough to take observations without catching our instruments on fire.

WARNING

Pro tip: Don't try to stare at the Sun without special observing glasses! The Sun's ultraviolet rays can easily burn your retinas and cause permanent damage.

REMEMBER

Also, not all observatories contain the same types of telescopes. Optical observatories use telescopes that see light in the infrared and visible portions of the EM spectrum, but millimeter-wave and radio observatories observe at longer wavelengths.

Telescopes that operate at different wavelengths look nothing like each other. For example

>> Optical reflecting telescopes have reflecting mirrors to capture light.

>> Optical refracting telescopes (used only in amateur astronomy today) are longer with two or more lenses connected via a tube.

>> Radio telescopes use the same technology as enormous satellite dishes.

TECHNICAL STUFF

Some radio telescope observatories are even larger by virtue of having dozens (or more) of radio telescope antennae. Their signals combine using a process called *interferometry*, a technique that increases the effective baseline of the telescope array to increase its sensitivity and detect smaller objects farther out in space. You can also do interferometry at visible wavelengths — the Large Binocular Telescope Observatory (see Figure 1-3) in Arizona has twin 28-foot (8.4-meter) mirrors and can combine the two beams of light to take observations of exoplanets and distant galaxies that would otherwise require a much larger single telescope.

TIP

Try holding your cell phone up to the eyepiece of a telescope. You're using the same principle that professional astronomers use with their enormous optical telescopes. Taking astronomical observations requires that the light from an optical telescope's mirror is directed into a scientific instrument. The two main types of instruments used at professional optical observatories are

FIGURE 1-3: The Large Binocular Telescope Observatory in Arizona.

>> An instrument that focuses light from an astronomical object into an image. This can be done with a specialized digital camera that's sensitive to minute variations in brightness, and sometimes combined with filters at multiple wavelengths. Observations through different filters can be combined to make color images, or ratioed to look for compositional differences and trends.

>> An instrument that splits astronomical light into its separate wavelengths. This can be done with an instrument called a spectrograph. When attached to the telescope, the spectrograph has a diffraction grating that spreads out the light into its individual wavelengths. Like a prism, this technique allows the spectrum of a star or galaxy to be recorded, providing information about its chemical composition.

Radio telescopes and other types of telescopes operate throughout the electromagnetic spectrum — check out Chapter 4 for more details.

Viewing from above: Space-based telescopes

Sometimes the top of a mountain just isn't tall enough to allow the observations astronomers need. Earth's atmosphere absorbs light from stars and galaxies at specific wavelengths of light, particularly at infrared and UV (and higher) portions of the EM spectrum. If the atmosphere is absorbing this light, it can't make it through to our telescopes. To avoid the Earth's atmosphere (who needs the atmosphere? only everything on the planet that breathes air!), astronomy must head upward into space and beyond the reach of Earth's atmosphere.

Placing a telescope into space requires launching it from a rocket on Earth. Think of a space-based telescope as a satellite that is also a telescope. A satellite in this context is any celestial body or piece of equipment that orbits the Earth. Space-based telescopes use specialized equipment to point toward a desired portion of the sky and record data. This data is then transmitted back to Earth using radio waves. Scientists analyze the data and may use it to create those famous images you've seen, for example, from the Hubble Space Telescope. Chapter 4 has more on space telescopes, and Chapter 18 contains a summary of 10 important space missions for astrophysics.

Stars, Galaxies, and Their Cosmological Friends

You're up to speed with how astrophysicists observe the sky, and have a baseline for the tools they use to make those observations. Next up: a brief tour of what's out there.

TIP

The objects visible from Earth in the night sky are at a huge range of distances, from close to very far away, but many of those objects are separated from you in time as well as in distance. For example, shooting stars — or meteors — are tiny flecks of cosmic dust that burn and glow in Earth's atmosphere as much as 100 km up. These are some of the closest objects to Earth. A larger space rock may make it down to the ground and land in your backyard as a meteorite, but that happens only rarely.

You may also see a satellite in the sky, perhaps the International Space Station or a communications satellite several hundred or thousand miles up. These human-made objects are orbiting the Earth and are farther away than shooting stars, because they're outside of the Earth's atmosphere. Also orbiting the Earth but even farther away is our Moon. It's about 239,000 miles (384,000 km) away from us but looks bigger and brighter than any star in the sky. Why? It's farther out past our atmosphere than a satellite but it's much larger; although the Moon is smaller than a planet or star, it's still significantly closer to us.

Beyond the Moon lie the planets of our solar system. Venus, Mars, Jupiter, and Saturn are the easiest to see due to their combinations of size and distance (Jupiter and Saturn are far but huge, and Mars and Venus are both near neighbors). Tiny Mercury is hard to see but can be seen with your eyes. Uranus and Neptune, however, are so far away that they require telescopes to catch. The Sun is the closest star to the Earth. We're 93 million miles (150 km), or 1 astronomical unit (AU), away from it, but the Sun is so big and bright that it dominates our daytime sky.

What about the rest of the stars that are so pervasive in the night sky? They're stars, just like our Sun, but farther away and appear dimmer. The closest star to us, besides our Sun, is Proxima Centauri at 25 trillion miles (40 trillion km) away. All of the individual stars in the sky are part of our Milky Way Galaxy. Our solar system is located in one arm of its spiral structure. With a telescope, you can see other faint and somewhat fuzzy objects in the sky. Some are nebulae within our own galaxy, but others are different galaxies, each of which can contain billions of stars.

Because light has a maximum speed (the speed of light, c, is 186,000 miles per second or 300,000 km/sec), light from these distant stars and galaxies can take thousands or even billions of Earth years to reach us. Cosmic distances like this are often measured in light-years — a light-year is the distance light can travel in one year through a vacuum, or about 5.9 trillion miles (9.5 trillion km). The distance to Proxima Centauri, for example, is about 4.3 light-years. It takes light 4.3 years to reach the Earth from Proxima Centauri, meaning that the light you see was actually given off 4.3 years ago. 4.3 years may not seem like much, but Proxima Centauri is relatively close. When you see distant galaxies, you're seeing light that was given off millions to even billions of years ago.

Every time you see pictures of a distant galaxy or star, you're looking back in time. Astrophysicists use these types of observations to peer through the cosmos all the way to the beginning of time itself, to the events surrounding the Big Bang. Between viewing ancient galaxies and detecting high-energy astronomical phenomena such as black holes and quasars, astrophysics works to solve a bit of a detective story of the universe.

Ready to dive into astrophysics? This book will take you on a ride through the universe, piecing together the elements (literally!) that make up stars, galaxies, and the universe — and, coincidentally, humans like yourself. Learn what you are seeing, how to see it, and what it all means, and then you'll be ready to learn about how the universe got here in the first place, and where it all might end.

IN THIS CHAPTER

» **Constructing the universe with fundamental particles**

» **Weighing the odds with matter and gravity**

» **Storing, using, and sharing energy**

» **Moving, grooving, and gravitating your way into understanding Newtonian motion**

» **Heating, cooling, and everything in between**

Chapter **2**

The A to Z of Physics

Where would you be without physics? Nowhere on this planet, that's for sure. Physics is a scientific description of how the world works on a fundamental level. Everything in the universe relates to four primary forces — gravity, the strong and weak nuclear forces, and electromagnetism. These forces not only unite tiny particles into atoms, but they also allow those atoms to stick together and form matter, stars, galaxies, cheese pizzas, and everything in between. Without these primary forces, we (and the universe) would cease to exist.

The earliest humans used physics in their daily lives. Constructing a cave dwelling, fashioning an arrowhead, and digging out a canoe all made heavy use of the basic principles of physics. The only difference today is that we've given these forces names and associated equations, and we are able to use what we know to model that which we cannot see or experience directly.

There's no astrophysics without physics. Before we get into the whole "astro" side of things, you need an understanding of the foundational physics. That said, compressing all of physics into one chapter of this book would be a meaningless undertaking. There's too much information to cover, and we wouldn't be able to do justice to any topic before moving on to the next. Also, although physics is grand and important and exciting (really!), not all of it is directly relevant to astrophysics. For these reasons, the scope of this chapter will be condensed to material that

>> Relates directly to topics covered in this book

>> Builds up the language and knowledge you'll need

>> Is fundamental to your understanding of astrophysics

From atoms to galaxy clusters, physics affects every detail of your journey into astrophysics. Let's begin!

Building Blocks of the Universe: Particles

We begin at the beginning. Or, put another way, let's start with the universe's tiny fundamental units: particles. Everything in the universe that we observe through a telescope or touch with our hands is made of matter. Whether it's a liquid, gas, star, or planet, matter is at the core, and matter is made of particles. Astrophysics, in a lot of ways, is the study of things that are huge and far away. It may seem strange to start out with something so small we can't see it with a microscope, but the fundamental particles that make up everything in the universe are generally the same particles that comprise the familiar stuff of our everyday lives here on Earth. Understand them here, and you are one step closer to understanding the cosmos.

WARNING

That "generally" in the previous paragraph is hiding the fact that the universe is really a mysterious place. There are many aspects, like dark matter and dark energy, that seem to be important components but may require new physics to understand. Or not . . . we don't know yet. For the meantime, we focus on the more familiar types of particles and cover the stranger elements in later chapters.

The big three: Protons, neutrons, and electrons

You're likely already familiar with the basic building blocks of matter as we know it. Protons, neutrons, and electrons are the backbone particles that comprise most of the ordinary matter in the universe. But, of course, particles by themselves won't get the job done. Combining particles into groupings called atoms allows these building blocks to start stacking into larger and larger objects and even molecules. Chemistry is the scientific field of study that focuses on this molecular level.

TECHNICAL STUFF

Atoms are made up of subatomic particles, the fundamental units that make up atoms. There are several main types of subatomic particles. We break this down a little: Table 2-1 has some basic properties of these fundamental particles, and Figure 2-1 shows the basic components of an atom.

TABLE 2-1

Basic Subatomic Particles

Type	Charge	Size
Proton	Positive	2000x mass of electron
Neutron	Neutral	2000x mass of electron
Electron	Negative	Almost no mass (2×10^{-30} pounds, or 9×10^{-31} kg)

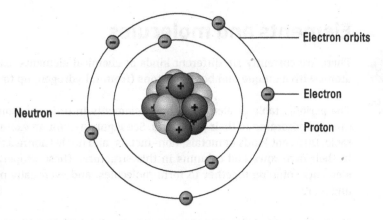

FIGURE 2-1:
The atom and its most basic components.

TIP

Atoms are the smallest building blocks of regular matter. Different kinds of atoms are made with different combinations of protons, neutrons, and electrons. In the right circumstances, atoms combine according to their charge and a set of rules that governs their structure. The simplest atom is hydrogen. In its standard form, hydrogen atoms consist of a single proton and neutron in the nucleus, with one orbiting electron.

TECHNICAL STUFF

In hydrogen, the positive charge of the single proton is balanced out by the negative charge of the single electron. The neutron itself has no charge, meaning that the net charge of a hydrogen atom is zero.

Atoms become more complicated when you start adding more protons and neutrons in the nucleus, balanced out by the same number of electrons orbiting the nucleus. Helium, for example, has two neutrons and two protons in the nucleus, and has two electrons in orbit. Carbon: six protons in its nucleus (and usually six neutrons), surrounded by six electrons. And so on.

The atomic number of an element is the number of protons in its nucleus. The atomic number uniquely identifies a type of element; hydrogen, atomic number = 1, will always have one proton, and carbon will always have six protons that give it an atomic number of 6. Atoms can gain or lose electrons through a process called ionization, and they can also gain or lose neutrons. An atom with a different number of neutrons than the base version is called an isotope. If the number of protons in an atom changes, it transforms into a different kind of element!

This model that has protons and neutrons in the nucleus and electrons orbiting, or in stacked energy levels, is called the classical model of atomic structure. Physicists now know that the real structure of atoms is a lot more complicated and involves probabilities rather than neat energy levels, but this simple description is useful in visualizing an atom's inner workings. It can successfully explain the many atomic and molecular interactions that make up chemistry.

Elements and molecules

There are currently 118 different kinds of chemical elements, each identified by atoms with a unique number of protons (from 1, hydrogen, up to 118, oganesson.

The periodic table of elements shows elements in order of atomic number and grouped according to their properties. See Figure 2-2 for an example of a periodic table. Different kinds of metals, non-metals, and noble (nonreactive) are grouped in their own rows and columns in this structure. These properties govern how elements combine together to form molecules, and eventually planets, nebulae, and more.

Most atoms are stable, and you can expect a hunk of iron to stay a hunk of iron over time. There are exceptions, however! We talk about the rare conditions under which one element can turn into another in:

>> Chapter 5 as we dive into nucleosynthesis in stars

>> Chapter 13 when we start approaching the Big Bang

That said, atoms of one element can bond together with atoms of another element to make molecules. Molecules are defined as atoms bonded together in groups in such a way that they comprise a new chemical compound. Figure 2-3 shows a molecule of water, typically consisting of two hydrogen atoms bonded to a single oxygen atom with the chemical formula H_2O. The central oxygen atom, with a nucleus of eight protons and eight neutrons, is surrounded by eight electrons in two different energy levels, as shown in a schematic form as concentric circles in Figure 2-3. The two hydrogen atoms are located at the top of the oxygen atom. Each hydrogen atom has a single proton in its nucleus and a single electron orbiting the nucleus, which bonds with the electrons orbiting the oxygen atom.

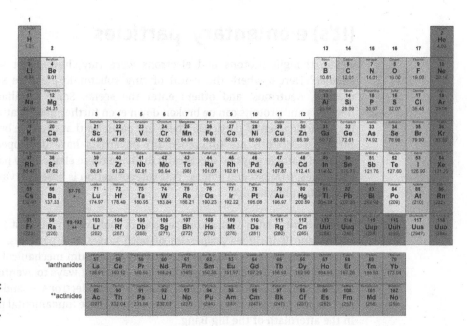

FIGURE 2-2: The periodic table of elements.

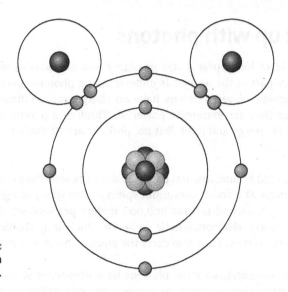

FIGURE 2-3: The structure of a water molecule.

In molecules such as water, the hydrogen and oxygen atoms keep their unique qualities. If the water molecules are split apart, the result is a return to simple hydrogens and oxygens. At the normal pressures and temperatures on the surface of the Earth, or even deep inside the Earth, the typical rules of chemistry apply and individual elements always retain their form. That's not always the case, and in astrophysics you'll encounter some decidedly odd environments. Sneak peek: The interiors of white dwarf stars are some of the oddest.

(It's) elementary, particles

TECHNICAL STUFF

If you thought protons and electrons were tiny, how about something even smaller? Here's where the world of tiny subatomic particles such as quarks, mesons, neutrinos, and others enter the scene. Some familiar particles like neutrons and protons can be broken down even further into what we call elementary or fundamental particles. These are particles that can't be broken down any further. (Electrons are fundamental!) Physicists have developed the standard model of particle physics specifically to divide these elementary particles into two categories, *fermions* and *bosons*. This categorization is based on various properties of quantum mechanics.

Quantum mechanics is a statistical way of looking at matter at its very smallest scales, and it includes properties of both waves and particles. The distinction between fermions and bosons is based on a quantum mechanical property called *spin*. All these tiny particles combine in a variety of ways to eventually form more familiar particles such as protons, neutrons, and electrons — and keep an eye out for Chapter 13, where you can learn about how these fundamental particles formed in the aftermath of the Big Bang.

Light it up with photons

TIP

It's impossible to look up at a star or planet and not wonder why it shines so brightly. A key part of the answer is understanding photons, particles consisting of electromagnetic waves. Photons have no charge and no discernable mass of their own, but they are extremely powerful. Think of a photon as a firefly — a small bundle of energy and light. But no, photons aren't made of light — light is made of photons!

REMEMBER

Photons are a kind of fundamental particle called a boson. They have no mass, and they always move at a fixed, maximum speed — the speed of light in a vacuum, 186,000 miles per second (or 300,000,000 meters per second). Photons are the particles that carry electromagnetic radiation, including visible light but also radio waves and x-rays. They also carry the electromagnetic force.

Like other fundamental particles, photons have properties of both waves and of particles. If you think of them as waves, you can define a wavelength that corresponds to the amount of energy carried by a photon — the shorter the wavelength, the higher the energy. We return to this idea later in this chapter when we go over the electromagnetic spectrum, a key concept in astrophysics.

CLEANER THAN A VACUUM

It's important to understand matter, particles, and energy in physics, but it's also important to think about the absence of those concepts. An area of space that has volume, but nothing in it, is called a vacuum. You've probably heard about the vacuum of space — the area between the stars and planets and galaxies is pretty empty, but not completely empty. There are some particles, energy, and radiation in just about everything, but their density varies a lot between the emptiest voids between galaxies and the thickest molecular clouds in spiral galaxies' disks.

An empty box here on Earth may look empty, but it's still filled with air. Air is not a vacuum; it's a bunch of particles in gas form. Even though you can't see those particles, they are indeed present. A vacuum can be simulated here on Earth in a laboratory using a vacuum chamber, a large self-contained volume that has been cleared of most air, matter, and particles to simulate conditions in space. Vacuum chambers are used to test space hardware that has to be able to operate in space, and also to perform delicate scientific experiments. Even the best vacuum chamber here on Earth, though, can't get close to simulating space. A super-good vacuum chamber on Earth might go down to a pressure of 10^{-10} Torr, a measure of pressure that corresponds to about 10^{12} particles per cubic meter. Deep space, however, has a particle density of closer to 6 particles per cubic meter!

What Matters About Matter?

Although many aspects of space are understood to the point where they can be defined and approximated — or, if you're really lucky, measured — other aspects are much less well-defined. Scientists might know, for example, that certain particles take up space and have mass, but it may not be clear what those particles are or where they came from. Enter the idea of "matter." This term applies to anything in the universe, terrestrial and otherwise, that occupies space and has mass.

Let's define a few basic terms:

>> **Mass** is the amount of matter in an object.

>> **Volume** is the mathematical description for the amount of space occupied by a three-dimensional object. It's easily calculated for shapes like boxes (length x width x height) or spheres ($4/3*\pi r^3$), but can also be calculated for planets! Our Earth, for example, has a volume of about 260 billion cubic miles (1 trillion cubic kilometers), and our Moon's volume is about 5.25 billion cubic miles (22 billion cubic kilometers).

>> **Density** is the amount of mass per unit of volume contained in an object or region.

GETTING HEAVY WITH MASS AND WEIGHT

You might hear people use the terms "mass" and "weight" interchangeably, but they'd be wrong. Weight is a measure of the gravitational force acting on an object. Mass, on the other hand, is a measure of the amount of matter an object has. Mass and weight are related by the formula $W = mg$, where W = weight, g = gravitational acceleration on Earth, and m = mass.

An object always has mass, whether it's out in space or sitting on the surface of the Earth, but its weight varies depending on the gravitational force to which it's subjected. Weight is always proportional to mass; an object on the surface of the Earth or the Moon with a large mass will have a higher weight than an object with a smaller mass.

REMEMBER

Note that density, mass, and volume aren't terms to be used interchangeably. Mass is how much matter an object has; volume is how much physical space that matter occupies; density is the amount of matter per unit of volume. Density is defined as mass over volume, $D = m/V$. High-density objects may be high in mass, but low in volume.

The water we drink and the air we breathe, or elements and molecules

Is everything in the world made of matter? One easy way to remember what matters with matter (ha!) is that objects with matter are tangible in some way. If you can hold it or taste it, there's matter. Air is made of matter, for example — it occupies space and has mass. Water is also matter — it's composed of different elements (hydrogen and oxygen, hence the H_2O formula), has weight, and takes up space.

To figure out what does not contain matter, think about these key indicators:

>> Does it take up space?

>> Does it have mass?

The light from a flashlight, for example, is not matter; it doesn't take up space or have weight. Sound is another good example of something that is not matter. Sound waves (and microwaves and other types of waves) also aren't matter!

Matter is made up of atoms and molecules; these combine into elements and compounds in different and exciting ways. Just a few of these combinations include humans, stars, and bananas. Most matter can be observed by the radiation it gives off or reflects. A special kind of matter called *dark matter* seems to have mass and take up space but doesn't interact with visible light or any other kinds of radiation — or even the electromagnetic force!

TIP

We talk more about dark matter, an important potential component of galaxies, in Chapters 11 and 15; everything else in this chapter pertains to "normal," non-dark matter.

It's all a state of mind with states of matter

TIP

To help understand a bit about the different kinds of matter, let's break them into four main types, or states. A "state of matter" is what scientists call the different forms the matter can take, because matter exists in multiple ways depending on the composition of its molecules and atoms.

>> **Solid:** Closely packed molecules, dense and firm(ish) to the touch

>> **Liquid:** Molecules in motion; takes the shape of whatever vessel it fills

>> **Gas:** Loosely dispersed atoms, very easily compressed, readily leaks out of an unsealed vessel

>> **Plasma:** Similar to gas but superheated and electrically charged

Matter can generally be transformed from solid to liquid to gas while still keeping its chemical structure and properties. These transformations are called phase changes. Liquid water, for example, can turn into solid ice or gaseous water vapor, but it will still have the same chemical formula of H_2O and the same chemical properties. If you take liquid water, freeze it, then thaw it again, you'll still have water. Elements can become plasmas, but many molecules will break up into separate elements when heated before they become plasma.

TECHNICAL STUFF

A phase change is different from a chemical reaction. Reactions are scenarios where two different kinds of atoms or molecules interact with each other to produce a completely different substance. An example of a simple chemical reaction is the burning of methane. In this reaction, methane gas and oxygen gas react with each other to produce carbon dioxide gas and water vapor. This reaction can be written as

$$CH_4 + 2O_2 \rightarrow CO_2 + 2H_2$$

If you count the number of carbon, hydrogen, and oxygen atoms on each side of this reaction, you'll see that the two sides are balanced — the 2 in front of the O_2 and H_2O helps ensure that you have the same number of each type of atom on each side. These sorts of simple chemical reactions involve rearranging each molecule's constituent elements into new configurations. After the reaction takes place, however, two new materials exist and these can't always convert back to their starting point (unlike with a phase change).

But sometimes it just doesn't (anti) matter

Matter is anything that has volume and mass. Antimatter is anything that takes up volume and mass.

Wait a second. So how are they different?

Most normal matter that we interact with every day is just that: matter. Antimatter is defined as its opposite. Every particle has an antimatter twin that has essentially the same mass, spin, and other properties . . . except for charge, which is the opposite. For example, a positron is a particle of antimatter, exactly like an electron, except that it has a positive charge.

WARNING

Most of the fundamental particles we talked about earlier in this chapter have an antimatter counterpart. If twinned matter and antimatter meet, however, they destroy each other and give off energy — in other words, don't expect a meeting with your evil twin to go well. Fortunately for us, antimatter is pretty rare in our universe, though it's possible that a parallel universe exists out there that's made up of our antimatter selves. (Also, antimatter passes through you constantly and it really isn't as scary as science fiction might have made you believe.)

Why is our universe made out of matter instead of antimatter? Some astrophysicists think it's a random circumstance following the Big Bang. In those early moments of the universe, matter and antimatter were created in approximately equal amounts, and annihilated each other. There was, however, a slight surplus of regular matter that was left over to form the universe. More on this in Chapter 13, but for now, let's go back to regular matter.

Let the Force(s) Be With You

There's no substitute for force, and it's seen constantly in our everyday lives. Drop a pencil off your desk, and the force of gravity will pull it down to the floor and keep it there. Flip the light switch and electricity, courtesy of the electromagnetic force, travels through the wires of your house to illuminate your lightbulb.

Here's a quick look at the essential forces of nature (and this book), in Table 2-2:

TABLE 2-2   ## Forces of Nature

Force	What It Does	Relative Strength	How Far Does It Extend?
Gravitational force	Force of attraction between bodies	Weakest of all 4 forces	Can act over incredibly long distances, but decreases with the square of the distance
Strong force	Force that binds atomic and subatomic particles	Strongest of all 4 forces	Most effective at atomic level, dissipates to zero over the length scale of a proton (10^{-15} m)
Weak force	Force that works on subatomic particles	Stronger than gravity but weaker than the strong force	Shorter distances than the strong force, dissipates to zero over a length scale of 10^{-18} m
Electromagnetic force	Force created when a magnetic field produces an electric field	Stronger than gravity	Decreases as one over the square of the distance

From Table 2-2, these four fundamental forces can be split up into pairs. The strong and weak forces are often called nuclear forces because they are incredibly important at the scale of an atomic nucleus, but are not experienced in our macroscopic world. Humans (as well as planets, stars, and galaxies) operate in a world that's governed by the larger-scale forces of gravity and electromagnetism. The matter that makes up those stars and planets and galaxies is held together by — you guessed it — the nuclear forces.

TECHNICAL STUFF

Physicists like to try to impose order on nature, and there's a desire to find a grand unified theory uniting all of these forces together. So far, electricity and magnetism have been united into electromagnetism, and that has been subsequently united with the weak nuclear force to form the electroweak force. There are models that would unite the electroweak force with the strong nuclear force, but those haven't yet been proven experimentally. Gravity has proved the most difficult force to try to unite. It's thought that all these forces were united for a brief, super–high-energy moment in time when the Big Bang occurred; more details are coming in Chapter 13.

Getting heavy with gravity

Imagine a world without gravity, the force of attraction between two bodies. If somehow gravity were to cease on Earth, the casualties would be endless. Not only would we float off the ground, but everything else would too — houses, cars, and everything around you would levitate.

WARNING

And by everything, we really mean everything — bodies of water (lakes and rivers, but also oceans) would have no reason to stay put. Neither would our atmosphere, so there goes breathing. And be careful what you wish for — more gravity isn't better than less in this case. Stronger gravity would pull everything in toward the center of the Earth, so most of our built environment would end in collapse.

Gravity in space is responsible for interactions big and small. As one example, consider a spaceship traveling from the Earth to the Moon. The ship needs a massive rocket to launch off the surface of the Earth. Launching requires overcoming the Earth's gravitational attraction, the force that wants to keep the rocket firmly here on Earth. After our rocket reaches space, it needs less and less power from the engine to counteract Earth's gravity.

TECHNICAL STUFF

As the rocket approaches the Moon, it's captured by the Moon's gravity and can then fire its rockets to come in for a soft landing. The engine used to land on the Moon (and hopefully blast off safely again) is much smaller than the engine required to blast off from the Earth. Why? Because the mass of the Moon is much less than the Earth, its gravitational attraction is less; our spacecraft needs less power to overcome it than when leaving Earth.

Both on Earth and in space, the force of gravity correlates directly to the masses of the objects involved and the distance between them. Seventeenth-century English physicist Isaac Newton developed the theory of universal gravitation to add mathematical specificity to this relationship, and it's identified by this equation:

$$F = \frac{Gm_1m_2}{r^2}$$

TECHNICAL STUFF

In this equation, F is the gravitational force between two objects. G is the gravitational constant, m_1 and m_2 are the masses of the two objects, and r is the distance between them. You can see from this equation that the force increases with the mass of the objects, and it decreases with the square of the distance between them.

A few final reminders about gravity:

>> Gravity acts on the gravitational centers of objects: Your center is pulled toward the Earth's center!

>> Gravitational collapse occurs when the inward pull of gravity is great enough to cause a cloud of gas and dust supported by gas pressure to collapse into itself and form a solar system.

>> Gravitational fragmentation occurs when a molecular cloud collapses into many different pieces that evolve separately. This can happen at different scales with pieces of cloud pulling away to form a star cluster, or a piece of a forming star system fragmenting into a planet around a new star.

>> Gravitational potential energy (GPE) is the energy a celestial object has due to its mass and position compared to another mass. It's described by $U = mgh$ where U = gravitational potential energy, h = height, g = gravity, and m = mass.

Thank Maxwell for the electromagnetic force

TECHNICAL STUFF

Like gravity, electromagnetism is a force that comes from a fundamental property of matter. Gravity is caused by objects having mass, and electromagnetism is caused by objects having charge. Initially, electricity and magnetism were considered to be two separate forces; objects could have either a positive or negative electric charge, and could have a positive or negative magnetic pull. In 1873, however, Scottish physicist James Clerk Maxwell published a theory that combined magnetism and electricity into the electromagnetic force.

TIP

Electromagnetism is created by the motion of charge within a conductor. This motion, the flow of electric current, then creates a magnetic field. Unlike gravity, which is only an attractive force, electromagnetism can either attract or repel, as there are both positive and negative charges. Two positive charges or two negative charges will repel each other, whereas a positive and a negative will attract each other.

The strength of the electromagnetic force, similarly to Newton's law of gravitation, is called Coulomb's law (named after French physicist Charles-Augustin de Coulomb) and it can be written as

$$F = k_e \frac{q_1 q_2}{r^2}$$

In this equation, F is the electromagnetic force, k_e is the Coulomb constant, q_1 and q_2 are the two charges, and r is the distance between them. The force is proportional to the size of the charges, and it is inversely proportional to the square of the distance; the closer the particles are to each other, the greater the electromagnetic force.

Electromagnetism is generally modeled as the flow of electrons through a conductor or through space; electrons are the carriers of the electromagnetic field. Maxwell came up with a series of equations (eponymously named Maxwell's equations) that govern the behavior of electromagnetic radiation. A solution to his equations in free space shows that electromagnetic waves move at the speed of light in a vacuum. This is true for all electromagnetic radiation, including visible light, x-rays, and radio waves. We talk more about the electromagnetic spectrum later in this chapter.

Strong and weak nuclear forces

The last two fundamental forces, the strong and weak nuclear forces, are responsible for the initial creation of particles and matter in the flash of time following the Big Bang. The strong force came into play when quarks united to create protons and neutrons, whereas the weak force used fusion to create the universe's first atoms. Although these two forces don't get nearly as much attention as gravity, they are absolutely essential at a subatomic level.

REMEMBER

Recall that the electromagnetic force causes two positively charged particles to repel each other. Sounds great unless you're trying to form atoms, because the nucleus of an atom usually contains multiple positively charged protons. Fortunately, the strong nuclear force, acting over very small atomic length scales, overcomes this electromagnetic repulsion and causes the protons and neutrons in a nucleus to stick together tightly.

The weak nuclear force is responsible for nuclear fusion. It operates only over very small atomic length scales, and it can cause a process called nuclear decay that changes a neutron into a proton. The strong and weak nuclear forces together are then responsible for

>> Nuclear fusion, when two smaller atoms join to create a new larger one

>> Nuclear fission, when a large nucleus breaks down into two smaller ones

Both of these processes can only take place at extremely high temperatures and densities, such as those found in the interior of a star or in a nuclear reactor. These processes release very high energy particles, which can be studied by astrophysicists.

Store It or Use It, But Don't Waste Energy

When you're low on energy, you might have a hard time getting work done. Energy in the physics world is technically defined as the ability to exert force on an object, or the ability of an object to perform work. Energy is the ultimate multitasker and comes in a variety of forms:

>> **Thermal energy** is energy generated by a heat source, including geothermally (from the Earth's core).

>> **Elastic energy** is stored via tension; the classic example is a wind-up toy or a squashed spring.

>> **Electrical energy** is created by electrons dashing through a conduit.

>> **Gravitational energy** is based on a body's height, or distance from its source of gravity; coast a skateboard down a hill, and that increasing speed is thanks to gravity.

>> **Chemical energy** is created and stored in the bonds between atoms in molecules; examples include natural gas and coal.

>> **Radiant energy** comes in waves of electromagnetic energy; it can include radio waves, gamma rays, and visible light.

>> **Nuclear energy** resides in an atomic nucleus, and is released in different fission or fusion processes depending on the atom.

>> **Sound energy** is created by waves that vibrate matter.

>> **Motion energy or kinetic energy** is in all rapidly moving objects and increases as the speed increases.

Fall faster into kinetic and potential energy

Do you feel like running, or sitting? Both use energy, though very different kinds. Anything in motion utilizes kinetic energy, a word that stems from the Greek *kinētikos* (motion). Throwing a ball, kicking a can, or riding a bicycle are all great examples of kinetic energy.

TECHNICAL STUFF

Potential energy is stored up by virtue of an object's situation, and is energy that can be maintained for later use. Raise a heavy ball over your head (but don't drop it on your foot!) — that ball has gravitational potential, or stored, energy. So does a boulder on the edge of a cliff. Figure 2-4 illustrates kinetic versus potential energy. Kinetic and gravitational potential energy are very useful later on in this book, when you learn about the formation of stars and how matter falls into a black hole.

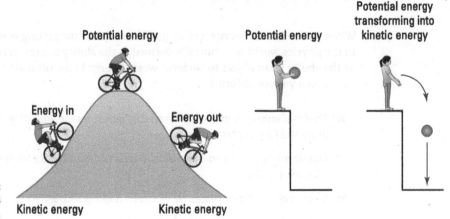

Forms of Energy

Potential energy

Energy in

Energy out

Kinetic energy

Kinetic energy

Potential energy

Potential energy transforming into kinetic energy

FIGURE 2-4: Kinetic versus potential energy.

Transferring thermal energy with conduction, convection, and radiation

Heat may not be a kind of matter, but it's certainly a type of energy. Specifically, heat is a measure of thermal energy, one that's involved in every aspect of astrophysics. Heat is generated when temperature excites molecules and atoms within a substance, causing them to collide and create heat. One major source of heat energy is a star — the Sun, for example, has a surface temperature of about 10,000°F (5600°C)!

There are three main ways in which heat is transferred between bodies and spaces:

» **Conduction:** Energy is transferred via direct contact (individual atomic motions)

» **Convection:** Energy is transferred through the bulk motion of material in a gas or liquid, such a lava lamp or hot oil.

» **Radiation:** Energy is transferred via electromagnetic waves, such as a heat light.

Deep inside a star, nuclear fusion takes place as hydrogen is turned into helium. This process releases vast amounts of energy. Different parts of stars transport energy in different ways, but in stellar atmospheres, it's primarily transported through convection. Warmer gas from down deep rises up to the surface, where it radiates some of its heat through the colder surface layers and out into space. The now-cooler gas then sinks back down to the hotter regions in the star's interior, where it heats up and rises again. After the energy leaves the star, it is transported through space by radiation, and this is the only type of energy transfer that works in the vacuum of space. See Chapter 5 for more about how stars create and transport energy.

More than the eye can see: The electromagnetic spectrum

TECHNICAL STUFF

And speaking of electromagnetic (EM) radiation . . . recall that radiation is a form of energy, one with magnetic and electric fields. EM radiation travels in waves that vary in length; those wavelengths determine whether those waves are infrared, visual, or something else entirely. Scientists use the EM spectrum to categorize waves based on wavelength.

REMEMBER

Remember from Chapter 1 that the Electromagnetic Spectrum (Figure 1-1) is a way of categorizing different types of electromagnetic radiation by wavelength. The spectrum covers everything from long wavelength to short wavelength, starting with radio waves, to microwaves, to visible light, to ultraviolet, x-rays, and gamma rays. With decreasing wavelength comes increasing energy; short-wavelength gamma rays are much more energetic than long-wavelength radio waves.

Astrophysical objects emit signals in the form of EM radiation, and astrophysicists use different types of telescopes to detect each kind of radiation. You are probably familiar with optical telescopes that detect light in the visible part of the spectrum; these use mirrors and/or lenses. Telescopes that gather IR or radio wavelength signals look like big satellite dishes. Telescopes that detect shorter-wavelength UV or gamma ray signals need to be launched above the Earth's atmosphere on balloons or satellites, because our atmosphere protects us from these wavelengths down here on the ground.

Light's unending flexibility as both wave and particle

Light is particularly fascinating to astronomers because it's one of the key ways in which the cosmos can be observed. The visible light illuminating the darkness is energy, and it's carried by photons (tiny massless particles that are also EM radiation waves). Light can be considered a particle because it's composed of photon particles, but light is also a wave in that it's propagated out to the world via EM radiation. How is this possible, you ask? It's both!

The actual description of light is derived from complex math in the field of quantum mechanics, an area that deals in complex statistical descriptions of interactions at the subatomic level. One way to think of light is as a complex probability distribution, but with particular values in its allowed energy properties. Consider a photon to be the smallest unit of light. Photons can act like waves because they bend and reflect but act like particles in that they can be defined with a particular fixed location and energy.

TECHNICAL STUFF

A theory called the Heisenberg uncertainty principle (after German physicist Werner Heisenberg) demonstrates the trade-off between the two. If you try to pin down the position of a photon, forcing it to act like a particle, then you lose certainty about its speed (which is a more wave-like property) and vice versa.

Some of the tricks of quantum mechanics, such as interference patterns, have experimentally verified light's wave-like and particle-like properties. Light can actually swap back and forth from one to the other over the course of a single experiment. Why, you ask? Because light is neither a wave nor a particle — remember, it's actually a complicated probability distribution. In fact, although quantum mechanics may be very difficult to understand on an intuitive level, it actually governs a lot of our modern technology. Computers, for example, are based on results from quantum theory. Although you may not understand exactly how your computer works, you accept that it does work (most of the time!).

It's the Law! (of Physics)

TIP

Physicists have been studying and writing rules based on observation of the environment since Aristotle's early treatises in the 4th century BCE, though Isaac Newton is credited with first formulating much of the basis for modern physics starting in the 17th century CE. In this sense, a law of physics is not a requirement that we humans place on the natural world, but rather a description of how the world behaves or a set of rules that we observe the universe following.

As physics has been formulated, many laws have been discovered and documented. Most are named after their discoverer or after the person who popularized them. Oftentimes, these laws of physics can be written in the form of a simple equation. Such laws typically start with simple, observable physical parameters such as mass, location, speed, and charge, and relate them to each other and to forces and energy values. There may also be a constant of proportionality that's used for conversion between different units.

Newton's laws of motion

If you've ever rolled a ball along the floor, you have an intuitive sense for how objects start and stop moving. The field of study governing this motion is generally called mechanics (not to be confused with the good folks who repair your car). The most fundamental basis for our understanding of planetary and astronomical motion lies in Newton's three laws of motion. These laws also apply to terrestrial life, of course, and facilitate our understanding of virtually all types of motion. Here are the key points to remember, summarized in Table 2-4:

TABLE 2-4   Newton's Laws of Motion

Newton's Laws	Remember As:	Description	Example
Newton's first law of motion	Law of inertia	Objects at rest tend to remain at rest; objects in motion tend to remain in motion unless acted upon by a force	A rolling ball will keep rolling unless it's intercepted, or stopped by friction
Newton's second law of motion	Force $F=ma$	Acceleration is reliant on a body's mass and the force applied to it	It takes more force to get a full wheelbarrow moving than it does an empty one
Newton's third law of motion	Action vs. reaction	Forces exerted on bodies are met by equal and opposite reactive forces	Birds can fly easily because as their wings push air away, air propels them forward

The basic idea of Newton's laws is that objects tend to keep doing what they are doing; if an object is moving, it will tend to stay moving, and an object that is stationary will tend to stay stationary unless acted upon by an external force. An acceleration is a change in motion and can be positive or negative, causing the velocity of an object to speed up or slow down. Newton's second law of motion relates mass and acceleration in the famous equation $F=ma$, where F = force, m = mass, and a = acceleration.

There are plenty of other key concepts in mechanics that we won't cover in detail in this book because they're generally less important to astrophysics; these include

>> **Friction:** A dragging force due to the impedance of motion

>> **Elasticity:** A property of matter which resists being deformed or stretched by an external force

>> **Work:** The transfer of energy to an object over a displacement

Step it up with velocity, acceleration, and momentum

WARNING

Life never stands still, and neither does physics. Dynamics in astrophysics is the general study of how objects in motion affect (and are affected by) four major properties: force, momentum, energy, and mass. Dynamics in this sense are not to be confused with dynamics in music! Composers and musicians use dynamics to describe how loud or soft a piece of music should be played, but these are unrelated to the mechanical motion of dynamics in physics.

One of the key aspects of force is motion, so let's look at a few very important concepts of motion in physics.

Velocity is the speed of any body in motion, referenced by its direction and indicated by v in equations. Its relationship to displacement and time is defined as

$$v = \frac{\Delta x}{\Delta t}$$

where Δx is the displacement, and Δt is the time interval. Acceleration is the vector rate of change of an object's velocity, indicated by a in equations. Its relationship to velocity and time is defined as

$$a = \frac{v - v_0}{t} = \frac{\Delta v}{\Delta t}$$

where v is the final velocity, v_0 is the starting velocity, and t is the time interval.

Momentum is a vector measurement of mass in motion, indicated by p in equations. Its relationship to mass and velocity is defined as

$$p = mv$$

where m is the mass and v is the velocity. Momentum and acceleration both pertain to a body's velocity, but differ in that momentum cares about an object's speed, whereas acceleration describes only how much an object's velocity is changing.

Having a conversation about conservation

And what is motion without limits? It's a fast trip to the emergency room, that's what. Our continued presence on the planet suggests that the world does not naturally want to destroy itself, and one of the ways in which life's forces remain in balance is through a few important conservation principles.

The law of momentum conservation mandates that if two objects collide in a closed system, the total momentum of both objects pre-collision is the same as their total momentum post-collision. In other words:

$$m_1 u_1 + m_2 u_2 = m_1 v_1 + m_2 v_2$$

REMEMBER

where m_1 and m_2 are the two masses, u_1 and u_2 are the initial velocities, and v_1 and v_2 are the final velocities. Remember here that momentum p = mass times velocity, or $p = mv$. In this equation, momentum can be transferred from one object to another in a collision. As a result, one object speeds up and the other slows down, but the total momentum stays the same. In physics terminology, momentum is conserved.

Similarly, the law of mass conservation requires that mass in a closed system does not change over time. The law of energy conservation means that in a closed system, energy is immutable; it can't be destroyed or created, but it can be transformed from other types of energy. This can be written as

$$K_1 + U_1 = K_2 + U_2$$

where K_1 is the initial kinetic energy, U_1 is the initial potential energy, K_2 is the final kinetic energy, and U_2 is the final potential energy. Collisions and conservation of energy will become important when you think about how stars and planets formed, first through small impacts in a cloud of gas and dust and then in larger and larger impacts.

TECHNICAL STUFF

In physics, a *closed system* is one where only energy can be exchanged between the system and its environment. An *open system* is one where both energy and matter can interact with their environment. An example of a closed system could be an insulated thermos of coffee, one that doesn't exchange heat with its surroundings and can keep your cup o' joe piping hot for hours. After you open the thermos and pour your coffee into a cup, however, the top of the cup can exchange heat with the outside world, and your coffee cools quickly.

We talked about momentum (p), defined for any moving object as $p = mv$, or a multiplier of the object's mass times its velocity. Angular momentum adds rotation to the equation and describes the relationship between the inertia of an object and its angular velocity. Angular momentum can be described mathematically for a circular orbit as $L = r \times mv$ where r = the object's position vector, m is its mass, and v is its velocity. More generally, for a non-circular orbit, the angular momentum $L = mvr \sin \theta$, where θ is the angle of motion. Angular momentum will become important in astrophysics when we talk about orbital motion, as well as the formation of stars and galaxies.

Angular momentum, like velocity and related quantities, is actually a vector quantity (one that has both magnitude [size] and direction). Vectors can be contrasted with scalar quantities, ones that have a size but no associated direction. The equations above should actually be performed as vectors using special multiplication procedures, but we've drilled them down to the essentials in the interest of comprehension and clarity. You're welcome.

Einstein and the mass energy equivalence

A final law of physics comes courtesy of German-American physicist Albert Einstein. Perhaps the most famous equation in the world of astrophysics is $E=mc^2$, also known as Einstein's mass-energy equivalence. We cover this concept in later sections of this book; in the current context, Einstein's discovery means a body's energy E changes in proportion to its mass m multiplied by the squared speed of light c.

Because the speed of light in a vacuum is a fixed but very large amount, if mass can be turned into energy, a small amount of mass can produce a huge amount of energy. This kind of reaction isn't all that relevant to our daily existence on Earth, but it's the basis for how the reactions at the centers of stars like the Sun give off so much heat and light.

Heat and Energy Unite with Thermodynamics and Statistical Mechanics

Thermodynamics is the study of work and energy within a system. It helps us understand the relationships between temperature/heat and work/energy, and how they transition between different forms as work and energy take place. Thermodynamics became a popular field of study in the mid-19th century with French engineer Nicolas Sadi Carnot's theoretical engine, and it became more fully developed as the steam engine rose as a means to generate power in the late 19th century.

Steam engines work by using a boiler to create steam. The expansion of that steam agitates a piston; the force created is connected to the crank and converted into rotational energy that can then be channeled into various uses. The basic principle behind a steam engine is that it converts heat energy into mechanical energy.

Why are we talking about steam engines? No, they aren't going to power your next starship, but they are a nice understandable starting point when you think about thermodynamics. Modern examples of thermodynamics are all around you. Home or office heating and air conditioning use thermodynamics, as do power plants (whether they're nuclear, thermal, or hydroelectric) and gasoline engines. And let's not forget the cosmos! The evolution of the universe is an excellent demonstration of the first law of thermodynamics. At the complete opposite end of the cosmic spectrum, a possible end-of-life scenario for the universe is called "heat death" (discussed in Chapter 16), and you'll soon be an expert in how this scenario is the ultimate triumph of the third law of thermodynamics.

TIP

To prepare you for the rest of this book, here's a very quick primer on units regarding temperature. Astrophysicists often use Kelvin, a metric system-derived unit that starts with zero Kelvin defined as absolute zero (the temperature at which all motion stops and the coldest temperature possible). This unit is named after Lord Kelvin, also known as William Thomson, a 19th-century British mathematician. By convention, the word *degrees* isn't used for Kelvin. From there, 0 Kelvin has a quick correlation to the more-familiar temperature scales of Fahrenheit and Celsius:

0 Kelvin = –459.67 degrees Fahrenheit = –273.15 degrees Celsius

Learn the laws of thermodynamics

There are four basic laws that describe the relationship in thermodynamics between heat, temperature, energy, and work in a closed system. We refer to various aspects of thermodynamics throughout this book, but here's one concise reference in Table 2-5.

TECHNICAL STUFF

Nature is full of tongue-twisters, but here's a good one: What's the difference between entropy and enthalpy in a thermodynamic system? Entropy is a measure of disorder within a system; specifically in thermodynamics, entropy measures the amount of thermal energy that's not doing work. Entropy in this context can be quantified as

$$\Delta S_{system} = \frac{q_{rev}}{T}$$

where ΔS_{system} is the total change in entropy, q_{rev} is the heat transferred in a reversible process, and T is temperature in Kelvin. This formulation works at constant temperature. The second law of thermodynamics states that energy in the universe tends to increase over time; because entropy is a measure of order, more disorder means higher entropy. The entropy of a gas is much greater than the entropy of the same substance as a liquid, which in turn is greater than the entropy of that substance as a solid.

TABLE 2-5 **Laws of Thermodynamics**

Law	Description	Real-World Example
0th law of thermodynamics	If any two systems are in thermodynamic equilibrium with a third system, they're in thermodynamic equilibrium with another.	If you put a hot cup of coffee B and a cold glass of milk C on a table in your room-temperature kitchen A and wait for a few hours, the coffee will cool down to room temperature and the milk will heat up to room temperature A — meaning that the coffee and milk are now in equilibrium with each other as well as with your kitchen.
1st law of thermodynamics	Total energy within a closed system can be transformed via heat and work, but cannot be destroyed or created.	Plant photosynthesis: Plants transform solar energy into chemical energy.
2nd law of thermodynamics	Entropy in the universe tends to increase over time; changes in entropy are always positive.	Adding ice cubes to a hot cup of water (ice melts and water cools as heat is transferred).
3rd law of thermodynamics	A perfect system at absolute 0 (0 K) has no entropy.	Predicted heat death of the universe — see Chapter 16.

Because ΔS tends to increase, if you open up a valve between a sphere that's full of gas and one that is empty, over time you would expect the particles to end up evenly distributed between the two spheres. This scenario represents a net increase in entropy because entropy is a measure of disorder. A system in which the particles are in one sphere and not in the other is a more ordered system, and so is lower entropy than the final situation in which the particles are all spread out evenly.

Enthalpy is the total amount of heat within the system, and can be written as $H = E + PV$. Enthalpy is usually written as H. It consists of the internal energy, E, plus the pressure P times the volume V. A particular chemical process can have a change in enthalpy associated with it, and in fact the enthalpy change of a chemical reaction is approximately equal to how much energy is gained or lost during the reaction. A chemical reaction is more likely to take place if the enthalpy of the system decreases during the reaction.

TECHNICAL STUFF

The Gibbs free energy equation describes how entropy and enthalpy are related. Together, they indicate the amount of usable energy in a thermodynamic system. The equation has a couple of forms, but the most common one is $G = H + TS$ where G is the total amount of Gibbs free energy, H is enthalpy (in joules), T is temperature (in K), and S is entropy (in joules/K).

Entropy becomes important as we talk about the end of the universe in Chapter 16. Because it's a measure of the total disorder of a system, and because entropy tends to increase over time, billions of years from now, the universe might come apart into an inert sea of broken-up particles. Stay tuned.

Truly being one with the environment: Blackbody radiation

How hot are stars? Scientists measure a star's temperature via blackbody radiation, also known as thermal electromagnetic radiation. All objects with temperature (that is, anything with $T >$ absolute zero, 0 K) give off electromagnetic radiation. With a few assumptions in place, you can directly correlate this radiation's peak in intensity with the temperature of the object. Take a good look at a candle flame; notice that the hottest parts near the wick have a bluish hue whereas the parts farther away appear yellow and red. The hottest parts with the shortest wavelengths look blue because blue light has a shorter wavelength. The cooler parts give off light at longer wavelengths, hence their yellow and red appearance.

How can you use this concept more scientifically? First, let's define a blackbody as a theoretical object that absorbs all the light that falls on it, and then emits that light over a broad spectrum of wavelengths. The blackbody will have a particular surface temperature. When you make a graph of the brightness of light given off versus the wavelength, as shown in Figure 2-5, you can create curves for different temperatures.

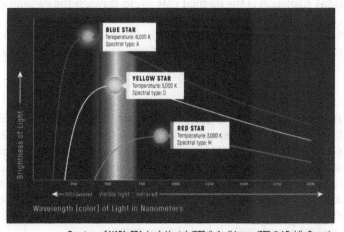

FIGURE 2-5:
Blackbody
radiation curve.

Courtesy of NASA, ESA, Leah Hustak (STScI), Andi James (STScI) / Public Domain

TECHNICAL STUFF

In this example, assume that the blackbody is a star. If the surface temperature of the star is 3000 K, the light will lie along the lower curve on the graph in Figure 2-5. If the surface temperature is 5000 K the star will lie on that middle curve, and if the temperature is around 8000 K, it'll exist on the upper curve. Because the temperature of the star can't be measured with a thermometer, you can plot its temperature versus brightness on this type of chart and look for the peak in brightness. This peak can then be used to determine the surface temperature and the star's color and spectral type (see Chapter 5 for more on stellar classifications).

TIP

Key takeaway: The peak on the blackbody spectrum gives a unique solution for the temperature of the star. It turns out that even though stars are not perfect blackbodies, using this technique provides a close approximation of the star's temperature. Blackbody assumptions are used whenever needed throughout this book.

Electromagnetic waves do the heavy lifting with radiative transfer

Consider the *therm* in *thermodynamics*. Stemming from the Greek word for heat, *therm* says it all — thermal shirts keep you warm in winter, thermometers give you a way to measure temperature, and a thermostat lets you control that temperature. The nature of heat is to radiate from its source, and radiation travels as electromagnetic waves. This concept, called *radiative transfer*, explains why we're able to experience warm sunlight; the heat generated at the core of a star (in this case, our Sun), travels through space and transfers heat to the Earth's surface.

WARNING

Without radiative transfer, the Sun's energy would never reach us because the other forms of heat transfer, conduction and convection, require some sort of material for the heat to move through (conduction) or to move the heat (convection). Radiation is the method that works through the vacuum of space.

Scientists initially used the Rayleigh-Jeans law to figure out the intensity of that blackbody radiation as a function of wavelength. It's defined by

$$B_\lambda(T) = \frac{2ck_BT}{\lambda^4}$$

Here, B_λ is the radiation's intensity written as the power emitted per wavelength lambda (λ). c is the speed of light, k_B is the Boltzmann constant, and T is the temperature. This equation works well at longer wavelengths; note that for shorter wavelengths, because the wavelength lambda is on the bottom of the equation, as lambda dwindles the intensity of the radiation B magnifies to the point where it approaches infinity (because you can't divide by zero!). This scenario is known as the Ultraviolet Catastrophe because it would potentially happen at the ultraviolet, or short-wavelength, side of the spectrum.

German physicist Max Planck went around this problem in 1900 by assuming that energy levels must be quantized. Energy had to be given off in discrete increments rather than being allowed to have any smoothly varying value, and the resulting step function is called Planck's constant. This is like having a clock whose second-hand ticks from second to second, rather than sweeping smoothly from one to the next through the intermediate fractional values.

Using this assumption leads to a different form for the intensity of black body radiation, one that correctly fits with our intuition that as the wavelength lambda approaches zero, the intensity of the radiation also approaches zero.

$$B_\lambda(\lambda, T) = \frac{2hc^2}{\lambda^5} \frac{1}{e^{hc/(\lambda k_B T)} - 1}$$

TECHNICAL STUFF

In the equation above, e (Euler's number) is an irrational number used in natural logarithms and other mathematical functions. It's similar to π in that it has a value that can't be written as a clean decimal or fraction.

Planck's discovery that energy levels were quantized was also used by Einstein when he developed the theory of the photoelectric effect, and this result eventually led to the development of quantum mechanics.

When the details matter, zoom in to quantum mechanics

TECHNICAL STUFF

A consequence of quantum mechanics is that you can know a particle's mass or velocity but not both. The details here would require a university-level textbook, but suffice it to say that quantum mechanics provides important solutions for various aspects of astrophysics. A few examples are

>> Quantum mechanics leads the way when determining what happens to matter and energy inside a black hole, and how Hawking radiation can escape its pull (see Chapter 8).

>> Quantum mechanics helps explain how subatomic particles were formed right after the Big Bang (see Chapter 13).

>> Quantum fluctuations right after the Big Bang likely govern the largest-scale distributions of matter in the universe: galaxy clusters and larger filaments (see Chapter 11).

>> Quantum tunneling and degeneracy pressure explain how white dwarfs and neutron stars can sustain their immense densities (see Chapter 8).

Although quantum mechanics may seem like the study of the very small, everything is connected in the field of astrophysics. Small-scale interactions can have wide-reaching consequences, and interactions and laws of physics that we can measure here on the Earth also govern physical properties in the cosmos, though with some notable exceptions. Now it's time to turn the page, strap in, and take a whirlwind tour through the field of astronomy!

Chapter **3**

Astronomy in a Nutshell

Physics is essential to understanding the mechanical, thermal, and quantum worlds around you. Similarly, astronomy is a critical stop on your journey into astrophysics because it launches you further into the night sky. Astronomers study our world beyond the atmosphere of the Earth, from bodies in our solar system to distant galaxy clusters that can only be seen with space-based telescopes.

Identifying those immediately visible landmarks in the night sky is arguably one of the most rewarding aspects of astronomy. Being able to look up and identify a few constellations, or understand that the white band in the sky is our very own Milky Way Galaxy, provides a sense of belonging and allows you to position yourself in the universe. Ancient scientists marked the position of stars and planets; navigators used the stars to find their way, and myths stemming from the night sky have sparked religious traditions that go back to our ancestral humans. Finding ways to make the universe understandable and applicable to ourselves has been part of human culture since our earliest days, and astronomy is the key to unlocking those mysteries.

We help begin your journey with an overview of astronomy that's most applicable to daily life. Learn how astronomers work, what astronomical measurements are all about, the details of observation, and the overall organization of the universe from planets to stars to galaxies. Why devote an entire chapter to astronomy in a book on astrophysics? The two are forever intertwined, both practically and academically. Astronomy is sometimes described as the "what," and astrophysics as the "why" or "how." This chapter takes you on a quick tour through the cosmos.

Where to Begin . . . Or, How It All Began

Humans have been keeping written records for about the last 5000 years. Ancient Sumerians used cuneiform symbols to track events in about 3000 BCE, and this concept spread to ancient Egypt and other civilizations. This process of recording history has been a part of cultures worldwide throughout the generations.

TIP

The first known documentation of astronomical events is found as early as 1300 BCE in China, and star parties were probably not top of mind. Astronomy was not (yet) a space race or an idle curiosity, but rather a practical necessity. Early farmers used the cosmos to track the seasons for planting purposes, sailors kept their bearings by using stars for navigation, and religious rituals relied heavily on stories generated from the night sky. These early sky watchers may not have known that the bright lights in the sky were distant suns like our own, but they observed and documented patterns in the sky to help understand the cycles of the seasons and the years.

TIP

Putting on your archeologist's hat for a moment, wind back the clock to 3rd century BCE Greece. Astronomers such as Ptolemy took astronomy from the realm of religion and observation into the start of a scientific field. Star and planetary movements were tracked with more rigor, though the mythology, art, and architecture dedicated to the Greek Pantheon (the 12 Olympian gods that governed ancient Greek culture) show that ties between astronomy and religion were still quite strong. By the 2nd century BCE, Hipparchus had developed the first star catalog using geometry to locate astronomical objects, and had created the first system for comparing the brightness of various stars via their magnitude.

During the Dark Ages in Europe, astronomy flourished in the Islamic world where astronomers made sophisticated observations between the 8th and 15th centuries CE. The ancient Mayan empire also developed both a calendar and an independent system of astronomical observations between 300 and 900 CE. European astronomical advances continued throughout the Renaissance. During this rebirth of

Classical themes, a flood of innovation was led by Polish astronomer Nicolaus Copernicus and Danish astronomer Tycho Brahe. In 1543 Copernicus went public with the heliocentric theory, which suggested that the planets orbited around the Sun instead of the Earth, and he essentially began modern astrophysics with a valid model of the orbits in the Solar System. Brahe's observatory and advanced star catalog further set the stage for today's knowledge of the cosmos.

The simplest tool for astronomical observation is one you carry with you at all times — your eyes. Depending on conditions, around 5000 stars may be visible with the unaided eye (though a portion is usually restricted by ambient lighting and atmosphere). For those times when you'd like a better view from your own backyard, a variety of telescopes are available to help magnify the cosmos. Telescopes, as popularized by Italian astronomer Galileo Galilei beginning in 1609, work by using curved lenses or mirrors to collect light, concentrate it, and magnify it. The larger the optics in your telescope, the more you'll be able to see.

Kicking it up a notch, observatories contain mega-versions of your backyard telescope. The twin Keck telescopes are located on Mauna Kea in Hawaii and have mirrors with diameters of 10 meters; these are some of the largest optical telescopes in the world. Radio telescopes, such as the former Arecibo Observatory in Puerto Rico or the Green Bank telescope in West Virginia, use huge dishes to collect radio waves that are then combined with amplifiers and receivers for recording and measurement.

Finally, space-based telescopes provide views deeper into space than anything we can view from Earth. These have the advantage of

>> A clear view of the heavens without any clouds or atmospheric turbulence blurring the scene

>> Hovering above the atmosphere, where space telescopes can observe wavelengths that are blocked by the atmosphere, such as ultraviolet and x-ray wavelengths

>> Observing 24 hours a day instead of having to wait for night

One of the most famous space telescopes, the Hubble Space Telescope (see Figure 3-1), was launched into orbit around the Earth by the Space Shuttle Discovery in 1990. More than 30 years later, it is still taking amazing observations of the cosmos; many of the gorgeous pictures in this book were taken by Hubble. Even though Hubble's mirror only has a diameter of 7.8 feet (2.4 meters), its location outside the blurring of Earth's atmosphere allows it to take unparalleled images.

FIGURE 3-1:
The Hubble Space
Telescope,
deployed into
space in 1990.

Courtesy of NASA

HELIOCENTRIC VERSUS GEOCENTRIC VIEW OF THE UNIVERSE

Although we all know someone who thinks they're the center of the universe, astrophysics would disagree. Going all the way back to the 2nd century CE, a geocentric model (the idea that the planets and Sun orbit the Earth) was the prevailing wisdom up until 16th-century astronomer Copernicus proposed a competing theory, the heliocentric model. This new theory stated that all planets, including Earth, orbited the Sun. This model was an elegant mathematical explanation that fit extremely well with decades of observations of the motions of the planets, and was confirmed when Galileo focused his newly updated telescope on the planet Jupiter and found unmistakable evidence that its four large moons orbited the planet, not the Sun.

Galileo's observations were not the first to promote the heliocentric theory. However, because his works were published in Italian, he came to the attention of the local authorities including the Catholic Church. The church considered the heliocentric model heresy, and placed Galileo under house arrest in 1633 until he recanted. He wasn't officially pardoned until 1992.

Mapping our Solar System, Galaxy, and the Universe

If it seems overwhelming to understand all the elements that make up the cosmos, you've come to the right place. Think of space in terms of scale. What's closest to you? What's farther away? Are some objects brighter than others, and are some consistently in the same positions relative to others? Humans have been trying, and mostly succeeding, to make sense of the universe for centuries. Fortunately, you can save a bit of time by starting at the end of this process.

Here's the short list of what you need to know about space, starting close to home and branching outward (as shown in the artistic view of Figure 3-2):

TECHNICAL STUFF

>> Our solar system, assuming you're also an Earthling, consists of the essentials: the Earth, our Sun, our planets (Jupiter, Saturn, Neptune, Uranus, Venus, Mars, Mercury), their moons, and various comets, asteroids, and dwarf planets.

>> Our Milky Way Galaxy contains our solar system, roughly 100 billion stars (many of which have their own array of planets and moons); nebulae; a variety of special types of objects including neutron stars, white dwarfs, and black holes; and two components that you will hear over and over in this book: cosmic dust and gas. Also called the interstellar medium, cosmic dust and gas consist of tiny particles combined with gas to create molecular clouds, which in turn are the foundational materials in stars, disks, and countless other astronomical bodies.

>> Our universe includes the Milky Way Galaxy plus as many as two hundred billion other galaxies (just in the observable part!), in addition to everything within them.

A question of scale

Have you ever gone to the dentist with what you're certain is a toothache, only to find out that you've got an underlying sinus infection that actually has nothing to do with your teeth? The human body is a system; understanding its individual parts is fundamental to your entire body's health, and each part of you (even your teeth!) provides valuable information about the rest. Gaining an understanding of the universe is no different. Studying planets in our solar system provides clues and information about our galaxy, nearby galaxies, and the entire universe.

Courtesy of NASA

FIGURE 3-2:
The solar system, galaxy, and universe.

But let's hold that thought and start with what's nearest to us. Our solar system contains eight planets including Earth, all of which orbit our Sun, and is divided into

>> The inner solar system: Mercury, Venus, Earth, and Mars. These are the terrestrial planets (also called the "rocky planets"), composed mainly of metal and rock.

>> The outer solar system: Jupiter, Saturn, Uranus, and Neptune. These are the "gas giant" planets, so named because they're composed of gasses such as helium and hydrogen. Uranus and Neptune are also sometimes called "ice giants" because they contain massive amounts of water vapor.

TIP

In addition to these planets, there are also a number of smaller bodies orbiting the Sun. The largest of these are the dwarf planets, a grouping that includes Ceres and Pluto (both former planets) as well as some of the other largest asteroids and Kuiper belt objects.

TECHNICAL STUFF

To understand astrophysics, it's essential to arm yourself with knowledge of scale. We typically use miles or kilometers as familiar units for measuring distance on Earth, but moving out into space requires a larger unit of measurement. Distances in the solar system are usually measured in Astronomical Units (AU). One AU is defined as 93 million miles (150 million km), the average distance from the Earth to the Sun. Distances beyond the solar system get even more, well, astronomical, and are usually measured in light-years. A light-year is the distance that light would travel in one year, or about 6 trillion miles (9.7 trillion kilometers).

So, using these numbers, consider distance:

>> The Earth's diameter is about 7910 miles (12,756 km).

>> The Earth's average distance to the Sun is 93 million miles (150 million km), or 1 AU.

>> Our solar system has a diameter of 178 billion miles (287 billion km).

>> The Milky Way Galaxy that our solar system resides in has a diameter of 105,700 light-years.

>> The galaxy closest to ours is the Canis Major Dwarf Galaxy, and it's a mere 25,000 light-years away.

>> There are at least 200 billion (yes, billion) other galaxies in the universe, and the objects farthest away from us are about 47 billion light-years from here.

>> The sum total diameter of the known universe is about 94 billion light-years.

The universe is only about 13.8 billion years old, and the speed of light is the fastest that anything can move. Given these facts, the sheer enormity of our known universe is evidence of cosmic expansion following the Big Bang, that point at which the universe started expanding out in all directions from a single point. We go into this subject in depth in Chapter 13, but for now, we start locally.

A planetary survey

Drum roll, please . . . we're finally ready to cover the subject of countless television shows, movies, theater productions, and even an entire symphony . . . the Planets. By which, of course, we mean the seven other planets in our solar system besides Earth. These would be Mercury, Venus, Mars, Jupiter, Saturn, Uranus, and Neptune. The chart in Table 3-1, details their relative size and temperature (in order by size), and we've included our Sun and Moon to make this a comprehensive quick reference. Figure 3-3 shows the planets to scale diameter-wise, though the distances to them are not to scale. We included our dwarf planet friend Pluto in this image, also to scale.

TIP

The planets orbit the Sun on orderly paths, and in the early 17th century, German mathematician Johannes Kepler (using data from Tycho Brahe, the astronomer who created precedents to modern-day star charts) derived three laws of planetary motion. These are shown in Table 3-2 and are illustrated in Figure 3-4.

TABLE 3-1   Solar System Details

Planet or Solar System Body	Diameter	Size Compared to Earth	Mean Surface Temperature
Earth	7926 miles (12,756 km)	YOU ARE HERE	59°F (15°C)
Sun	870,000 miles (1.4 million km)	109x Earth	10,000°F (5530°C)
Jupiter	88,846 miles (142,984 km)	11x Earth	–166°F (–110°C)
Saturn	74,898 miles (120,536 km)	9x Earth	–220°F (–140°C)
Uranus	31,763 miles (51,118 km)	4x Earth	–320°F (–195°C)
Neptune	30,607 miles (49,528 km)	4x Earth	–330°F (–200°C)
Venus	7521 miles (12,104 km)	0.95x Earth	867°F (464°C)
Mars	4220 miles (6,792 km)	0.53x Earth	–85°F (–65°C)
Mercury	3032 miles (4879 km)	0.38x Earth	333°F (167°C)
Moon	2159 miles (3475 km)	0.25x Earth	Ranges –298°F (–183°C) to 224°F (106°C)

FIGURE 3-3:
Diagram illustrating relative planetary size (with bonus Pluto).

Courtesy of NASA/Lunar and Planetary Institute

TABLE 3-2   Kepler's Laws of Planetary Motion

Law	Description
Kepler's first law	All planets in the solar system orbit the Sun in elliptical orbits, where the Sun is one of the ellipse's foci.
Kepler's second law	The Sun can be connected to other planets via a radius vector that is divided into equal regions, in equal time segments.
Kepler's third law	The square of a planet's orbital period is proportional to the cube of the elliptical orbit's semi-major axis. This can be written as: $P^2 \propto a^3$ (p = period of orbit in years, a = semi-major axis of elliptical orbit in AU).

Kepler's laws of planetary motion

First law

All planets move about the Sun in elliptical orbits, having the Sun as one of the foci.

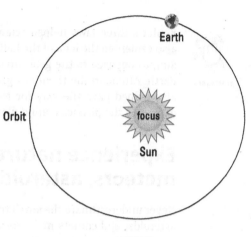

Earth

Orbit

focus

Sun

Second law

A line joining a satellite and the Sun sweeps out equal areas during equal intervals of time.

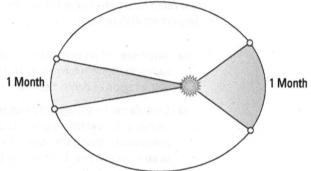

1 Month

1 Month

Third law

The square of a planet's orbital period is proportional to the cube of the elliptical orbit's semi-major axis.

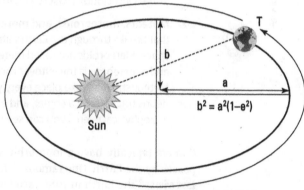

$$b^2 = a^2(1-e^2)$$

Sun

T

b

a

$$T^2 \propto a^3$$

T = Time to Complete Orbit
a = Length of Semi-major Axis

Kepler's Third Law of Planetary Motion

FIGURE 3-4:
Kepler's laws of motion.

REMEMBER

Kepler's three laws helped scientists make sense of the sky's movement. They also cemented the idea of the heliocentric solar system where the planets orbit the Sun, as opposed to the geocentric view that had the Sun and planets orbiting the Earth. Although the theory of gravity hadn't been formulated yet, Kepler's work also helped pave the way for Newton's law of gravitation (see Chapter 2) that explained the physics behind planetary motion.

Experience nature's light show with meteors, asteroids, and comets

Never underestimate the small things in life. Relatively small bodies like meteors, asteroids, and comets make for a stunning light show if you're lucky enough to see them in action. These three types of astronomical objects are similar, but have important differences:

>> **Asteroids** are primordial chunks of metal and rock that were created during early solar system formation. Many asteroids orbit the Sun in the asteroid belt, a region of debris between the orbits of Mars and Jupiter.

>> **Comets** can be thought of as ice balls orbiting the Sun. They consist of rock, cosmic dust, and frozen gas that, as they approach the Sun, partially sublimates, surrounding the comet nucleus with a cloud of gas. This gas can stretch out in a long tail that may be visible from Earth. Comets can originate from the Kuiper belt, a region of icy solar system leftovers that includes Pluto, or the Oort cloud, a vast region in the far reaches of the solar system.

TIP

>> **Meteors, meteoroids, and meteorites** are falling rocky and metallic debris that breaks through the Earth's atmosphere. A meteoroid is a piece of rock in space. Meteoroids are called meteors after they fall through the atmosphere, and a meteorite is the leftover rock after it reaches the surface of the Earth. Meteor showers take place every year as the Earth's orbit passes through the debris track left by a comet, and are one of the most visually stunning astrophysical events you can see unaided.

Comets typically have a long orbit around the Sun and can make multiple close passes by the Earth. One famous comet, Halley's comet, has a 76-year period and last visited the Earth in 1986 (see Figure 3-5). It will next be visible in 2061; plan accordingly for optimal viewing!

Galaxies and beyond

As we move farther from home, just as the Earth is part of our solar system, the Sun and its planets are part of the Milky Way Galaxy. The Milky Way is a spiral galaxy, and our solar system is located in one of the spiral arms, conveniently located far away from the supermassive black hole at the center.

Courtesy of NASA Jet Propulsion Laboratory, California Institute of Technology

FIGURE 3-5:
Halley's Comet,
last viewed from
Earth in 1986.

TIP

The disk of our own galaxy is visible in the sky as the Milky Way, that bright band of stars that stretches across the sky. You can also see other objects in the sky — a small telescope will reveal star clusters and fuzzy blobs of gas and dust called nebulae, another type of celestial object featured in the Milky Way Galaxy. Some of those fuzzy blobs are located far beyond our own galaxy! Chapter 9 goes into more detail both on our home galaxy and on other nearby galaxies.

Observational Astronomy: What are Those Dots in the Sky?

Observational astronomy helps you make sense of the heavens. Gaze up at the sky on a clear evening and you see a plethora of tiny, sparkling dots. Several thousand of those are stars; we take you on a voyage through the cosmos shortly. But is everything out there a star? The short answer is no, and here's why:

>> Seven of those twinkling dots are the planets in our solar system. Many are visible with the unaided eye, depending on your location and viewing conditions.

>> Comets are rare to witness and asteroids will require a telescope, but meteor showers (raining debris that appear as quick, brilliant flashes in the sky) are much more common. They're hard to mistake for stars, though, because you'll witness them in motion.

>> Are those blinking and often moving objects UFOs? Not so much, no. Most of them are either satellites, beacons, or moving (and definitely Earth-based) aircraft.

>> If you're really lucky, you might catch a glimpse of the International Space Station (ISS) as it travels from west to east. It doesn't blink as an aircraft would and doesn't appear to fall to Earth like a meteor.

REMEMBER

And now, back to the stars! That seemingly-random series of dots in the sky is actually quite orderly. Stars always appear in the same position relative to each other. The sky moves as the Earth rotates on its axis, but stars remain in the same positions and are arranged into patterns or groups called constellations. Ancient civilizations looked up at the night sky and saw the same stars that we do today, and that's perhaps the most humbling aspect of astronomy.

TIP

Every star you see has been seen by the generations that came before you. Let that sink in for a moment before moving on.

Early human civilizations realized that stars appeared in the same patterns every night. Cave paintings and carvings from thousands of years ago depict constellations, though the first written records of named constellations stem mainly from research by ancient Greek astronomer Ptolemy along with ancient Chinese observers. In the more modern era, the constellations were codified and used as a tool of science as well as navigation and timekeeping.

Constellations: Mapping the stars

TECHNICAL STUFF

The earliest defined constellations were named after shapes that they resembled, although the description varied from culture to culture. In 1922, the International Astronomical Union officially divided the sky into 88 defined constellations. Some are famous, like Orion the hunter, and some such as Corvus the crow are more obscure. Constellations are a way of dividing up the sky into known pieces of real estate, and a method of identifying stars and star systems.

TIP

Imagine drawing a giant sphere around the Earth and painting the stars on that sphere. This is the basic model used by ancient observers, and they devised a coordinate system that's still used today. To find a location on the surface of the Earth, we draw imaginary lines going north-south (longitude) and east-west (latitude) to make a three-dimensional grid on the surface of our (mostly) spherical Earth. The coordinates of every location on Earth are defined using latitude and longitude lines. This method is, incidentally, how the GPS in your car can help you find the nearest gas station.

Similarly, you can use the celestial sphere to find coordinates on the sky. These coordinates are called *right ascension* and *declination*. They're similar to the familiar grid of latitude and longitude, except that they're on the inside of a sphere facing inwards towards us. Two-dimensional coordinates like this can't account for the fact that stars actually are all at different distances from us, so this approximation can only pinpoint location and not distance. For comparison, on Earth if you have latitude and longitude, mountain heights can't be identified without the third aspect of altitude.

TIP

The 88 constellations defined by the International Astronomical Union have boundaries expressed in right ascension and declination, and any stars within those boundaries are part of the officially named constellation (no matter whether they fit in with the actual drawing of the constellation in any way.)

Because the Earth is a sphere, the entire night sky cannot be seen from your backyard. If you live in the southern hemisphere, you can see the southern half of the sky; those living in the northern hemisphere can witness the array of constellations in the northern sky. If you're lucky enough to live on the equator, however, you can see the whole sky over the course of the year, though still only half at a time!

Of the 88 total official constellations, 52 live in the southern hemisphere, whereas 36 are found in the northern. The entire night sky in the northern hemisphere appears to rotate around the north star, Polaris, a conveniently located beacon near the North celestial pole. Its location makes it particularly useful for navigating the northern hemisphere.

The ecliptic, a year-long journey of the Sun

The Sun isn't in a constellation, of course, but it does impact our favorite groupings of stars via the concept of the ecliptic. As the Earth orbits the Sun every 365 days, its elliptical path can be visualized as a plane. The ecliptic is this plane as projected onto the "celestial sphere," the imaginary globe placing Earth's core at the center. Because the Earth's axis is tilted, the angle of the ecliptic is also tilted with respect to the celestial equator — see Figure 3-6 for the details of how this geometry works.

TIP

Here's where the fields of astronomy and astrology come together. Don't worry, we aren't going to predict your behavior or future. That said, the signs of the Zodiac are actually 12 constellations located along the ecliptic, the path the Sun travels over the course of a year. Now you know the real meaning behind Capricorn or Taurus!

The ecliptic is also important because it represents not just the path of the Sun, but also the plane of the solar system as projected onto the sky. Remember how the constellations represent fixed patterns of stars that always have the same relation to each other? There are some celestial objects that break that mold, and the planets are among the most prominent. In fact, *planet* comes from a Greek root meaning *wanderer* because the planets seem to wander through the sky; they move from constellation to constellation, always along the ecliptic.

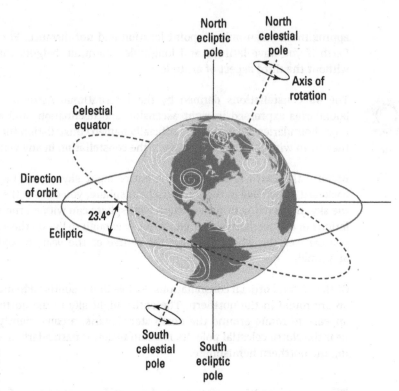

North ecliptic pole

North celestial pole

Axis of rotation

Celestial equator

Direction of orbit

23.4°

Ecliptic

South celestial pole

South ecliptic pole

FIGURE 3-6: Diagram of the ecliptic.

When the Moon crosses over the plane of the ecliptic, an eclipse occurs. Note how similar those two words sound? Eclipses occur only along the line of the ecliptic. Say that five times fast!

How astronomical objects are named

To misquote William Shakespeare, a star by any other name would sparkle just as bright. You're likely familiar with the most famous star but perhaps by a few names. Polaris, the Pole Star, the North Star — these are one and the same. How is a star's name determined? There are a few ways:

» Bright stars are named independently. Betelgeuse, for example, exists in the constellation Orion but it's bright enough to have warranted its own name. Many stars have Arabic names because they were first cataloged by ancient Islamic astronomers.

>> Stars can be named after the official constellation that they're in. 51 Pegasus, for example, is the 51st catalogued star in the Pegasus constellation.

>> Other stars only have catalog names. For example, HD114762 is the 114,762nd entry in the Henry Draper star catalog.

>> Still other stars are given catalog IDs via the SAO — Smithsonian Astrophysical Observatory Star Catalog. Many of these cataloged stars also have their own names. SAO 67174 is the bright star Vega, for example.

>> Sorry, one more: Still other stars' names start with Greek letters because using the Greek alphabet was the original system created by German astronomer Johann Bayer in 1603. For example, the brightest star in the Pictor constellation was named Alpha Pictoris (Pictoris is the possessive form of Pictor in Latin), the second star was Beta Pictoris, and so on. For constellations with more than 24 stars, Bayer then moved over to the Latin alphabet from A–Z.

If you think naming stars is complicated, how about naming the planets that orbit them? As astronomers started finding exoplanets orbiting stars beyond our own Sun, they had to come up with naming conventions for those planets (it would get confusing to call them all Bob, for example). See Chapter 7 for more info on the solutions that astronomers have come up with.

REMEMBER

From cosmic timescales and distances to the layout of the heavens, astronomy helps you understand your place in the universe. The physical organization of the solar system, the structure of the universe into galaxies, and the notion of fixed stars in constellations that are interrupted by the motions of the planets, all were essential to developing an understanding of all the different parts of the universe.

The next step beyond this descriptive understanding is a focus on why and how. How does a star give off so much heat and electromagnetic radiation? Why does a galaxy look the way it does? How did the universe form, and how will it end? Those are questions which move into the realm of astrophysics; their answers combine the physical models of Chapter 2 with the astronomical descriptions in this chapter. Ready or not, astrophysics is next!

IN THIS CHAPTER

» Understanding space from a physics perspective

» Choosing, using, and studying data from the cosmos

» Waving through the frequencies of the EM spectrum

» Radiating heat and more via heliophysics

» Don't look up! Taking careful advantage of eclipses

Chapter **4**

Bridging the Gap Between Astronomy and Physics

A stronomy, cosmology, astrophysics, planetary science, physics . . . the lines between the various fields of space science are blurred, but in a wonderful way. Astrophysics is a uniquely beautiful combination of astronomy and physics in that it applies the laws and constructs of physics to the knowledge of the universe gained through astronomy. At a high level, astrophysics encompasses some of the most complex concepts out there. From the origin of the universe to the formation and dynamics of stars and galaxies, astrophysics uses all of space as its testbed. We look at the major ideas required to understand astrophysics, as well as the most important ways in which astrophysics is conducted and how data is gathered.

Finally, we bring the discussion back home with two very important topics, helio-physics and eclipses. These are both examples of astrophysics within our own solar system. Although much of astrophysics looks to the world beyond the Milky Way, there's an incredible amount to learn from what we can see and experience with our own eyes. If you're going to study the stars, why not start by studying our very own?

More Than the Sum of Its Parts: The Unique Study of Astrophysics

You've made it this far, but what is astrophysics, anyway? Astrophysics is more than a subfield of either astronomy or physics. It requires uniquely skilled scientists who have a background in both but also have a yearning to answer the big questions:

>> How did we get here?

>> How did our stars evolve?

>> How did the universe begin, and how will it end?

If you're seeking an understanding of the inner workings of the universe and a safe space to look for answers to those big questions, you've come to the right place. Astrophysics delves into not just the "what" behind space, galaxies, and celestial objects, but also the "why" and the "how."

Although astronomy can focus on observations taken with telescopes here on Earth or in space, astrophysics focuses more on the interpretations of those observations. Astrophysics can involve theoretical investigations that make predictions that can then be tested through observation. It often requires the construction of bigger and better telescopes (or other detectors) to check those predictions and see whether they are proven wrong or right. Astrophysics can also take place in a laboratory, where physical processes are simulated under conditions similar to those found in deep space.

TECHNICAL
STUFF

Spectroscopy, for example, can be performed under laboratory conditions to study the compositional signature of known materials, and these results can then be compared to spectroscopic observations made by astronomers to help determine the chemical makeup of a distant star or nebula.

One key concept in astrophysics is distance, and one way that distance is measured is called the *redshift*. Redshift calculations are made using the same spectral signature that provides composition; these calculations then take advantage of some fundamental properties of physics to provide velocity, and then distance.

One shift, two shift, redshift, blueshift

When a child looks up into the night sky and tries to touch a star, they realize that they're too far away and ask you to put them on your shoulders. When you do and they reach out again, they've gained a new piece of information — stars are farther away than the biggest thing they know (you). All of astrophysics mimics this kind of detective story; the universe leaves us clues, and it's up to us to figure out how to interpret them.

REMEMBER

We refer to the detective story theme throughout this book as you navigate your way through the tiniest and grandest of astrophysical scales.

TECHNICAL STUFF

Fortunately, finding distances from here to the galaxies is a solvable problem. Scientists use the concept of redshift to understand distance, and here's how. First, remember that the electromagnetic spectrum is the full range of electromagnetic (EM) radiation (energy traveling at the speed of light). The visible part of the spectrum is the range of wavelengths discernible by the human eye, between 700 and 380 nanometers.

You might be familiar with a phenomenon known as the *Doppler effect*, a fundamental concept in physics that works on sound waves. When an object making a sound moves towards you, such as a train blowing its horn, the sound waves compress because the train is moving in the same direction that the sound wave is moving; the net effect is shortening the wavelength of the waves. This compression results in an increase in the waves' pitch, making the train's horn sound higher. Similarly, after the train moves past you, the train is now moving in the opposite direction from the sound waves traveling towards you. The wavelength gets stretched out, and the pitch of the train's horn goes down.

Although there's no way sound can travel through the vacuum of space, this effect also applies to light (with some numerical differences), and it's illustrated in Figure 4-1. When objects move away from you, the light they emit is stretched out and the wavelength becomes longer as it shifts towards the red part of the EM spectrum. This is called a redshift. Similarly, when objects move toward you, the light wavelengths become shortened and are shifted towards the blue part of the spectrum; hello, blueshift!

The amount of redshift or blueshift can be used to determine how fast an object is moving. Each object gives off a unique signature of spectral lines. These are caused by absorption and emission of light related to the chemical makeup of the object — we talk more about this concept in Chapter 6.

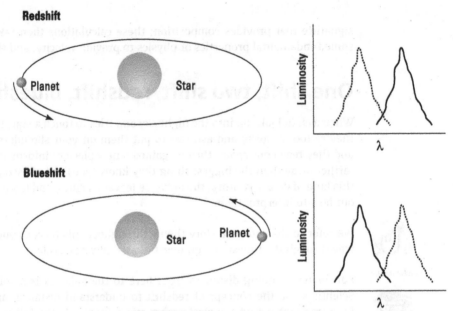

Redshift

Planet

Star

Blueshift

Star

Planet

Luminosity

λ

Luminosity

λ

FIGURE 4-1:
Redshift versus
blueshift.

TIP

Spectroscopy offers patterns of lines with known separations from each other. Although it would be impossible to prove the expected identity of a single spectral absorption line of a star or galaxy, the pattern of lines of each element is unique enough that you can slide it up and down the spectrum to determine its offset. By shifting the lines of many elements together, astrophysics can figure out the velocity and composition of a star or galaxy. The direction of offset tells you if the shift is a redshift or a blueshift, and the amount of the offset is proportional to the velocity of the object.

Using the Doppler effect to determine an object's velocity is of clear benefit to astrophysicists, but American astronomer Edwin Hubble went a step further when he discovered a relationship between redshift and distance. That discovery, explained in more detail in Chapter 9, was foundational to the realization that some fuzzy spots of light in the sky were actually galaxies far beyond our own that shine in an expanding universe.

Don't be late to the party
when time matters

REMEMBER

Although distance is a key concept in astrophysics, a second major area of focus is time. We've looked at electromagnetic radiation, and explained how the speed of light is a constant. Light travels incredibly fast, but the universe is full of distances that are almost impossible to comprehend. When you look up at the sky, you are seeing light that has traveled through space from the object you are seeing, all the way back to you. Because the light takes a finite amount of time to reach you, you are actually seeing the image of an object as it appeared when that light was given off — that means, in the past!

In our own solar system, the Sun is about 93 million miles (150 million km) away from the Earth, and it takes light about eight minutes to travel that distance. When you look up at the Sun with filters (if you value your eyesight, please don't stare into the Sun without proper eye protection), you are seeing the Sun as it appeared eight minutes ago. If you want to see what the Sun is doing *right now*, you'll have to wait another eight minutes for that light to make it back to us on Earth.

Is that particular delay problematic? Not so much: Eight minutes isn't that long in the grand scheme of things, and the Sun doesn't change much on that sort of short timescale. However, as cosmic distances increase, the time delay gets larger and larger.

TECHNICAL STUFF

These distances are often measured in light-years, the distance light can travel in a year. One light-year is 6 trillion miles (or 9.7 trillion kilometers). Figure 4-2 shows a few cosmic distances in light-years. These are noted from the closest star (Proxima Centauri), only 4.3 light-years from the Earth, out to an ancient galaxy that's 13.4 billion light-years away!

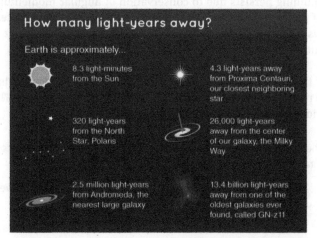

How many light-years away?

Earth is approximately...

8.3 light-minutes from the Sun

4.3 light-years away from Proxima Centauri, our closest neighboring star

320 light-years from the North Star, Polaris

26,000 light-years away from the center of our galaxy, the Milky Way

2.5 million light-years from Andromeda, the nearest large galaxy

13.4 billion light-years away from one of the oldest galaxies ever found, called GN-z11

FIGURE 4-2: Cosmic distances, in light-years.

Courtesy of NASA / Public Domain

WARNING

Those cosmic distances don't just mean that it would take us a really long time to travel there, even if we invented a spaceship that could travel close to the speed of light (spoiler: warp speed hasn't been invented yet!). They also suggest the amount of time it takes the light from those objects to reach us. When we look at

Proxima Centauri, we are seeing it as it looked 4.3 years ago. When we look at the Andromeda Galaxy, we see it as it appeared 2.5 million years ago. And when we look at an ancient galaxy like GN-Z11, it's as if it appeared 13.4 billion years ago.

Looking at distant objects is looking back in time, and this simple fact allows astrophysicists to study the history of the universe just by looking up at the sky and observing objects that are farther and farther away. Galaxies like GN-Z11 must have formed soon after the universe itself formed 13.8 billion years ago, and observations of objects like these tell us about conditions soon after the Big Bang. See Chapter 13 for more about how we think the universe began, and how those observations are used.

Celestial mechanics and orbits

Although light is a super-important part of understanding the cosmos (because without it, we'd all be in the dark!), when time and distance are involved, motion is also a key player. Celestial mechanics applies concepts of physics to the motion of bodies in space. The Earth is constantly spinning about its axis, as well as orbiting our Sun, thanks to the constant forces exerted by gravity. And it's not just the Earth — moons orbit planets, planets orbit stars, and stars orbit the centers of their galaxies.

That spinning persists due to conservation of angular momentum, a basic principle of physics stating that the rotational spin motion of a system persists until that system is acted upon by an outside torque. See Chapter 2 for more about momentum.

One of the most important rotational concepts in astrophysics is "orbit," or the curved path a body takes around its host moon, planet, or star. There are two key definitions here around orbits and rotation:

>> The **rotational period** for a system is the time for an object to complete one rotation around its own axis.

- The Earth's rotational period, for example, is 24 hours (relative to the Sun) because it completes one rotation per day (by definition).

- The Sun's rotational period around its own axis is about 27 days (relative to the Earth).

>> The **orbital period** for a system is the time for one body to make a complete orbit around its host moon, planet, or star.

- The orbital period of the Earth, for example, is 365 days or one year (by definition).

- The orbital period of our solar system around the center of the Milky Way Galaxy, our home galaxy, is 225 million years, or one cosmic year (by definition).

TECHNICAL STUFF

Planetary orbits are elliptical (and circles are a special kind of ellipse), thanks to gravity and Kepler's laws of motion (flip to Chapter 3 for a quick refresher), which describe this motion. The Earth's orbit around the Sun is an elliptical path with the Sun at one focus of the ellipse; ergo, the distance between the Sun and Earth varies over the course of a year. As seen in Figure 4-3, the point at which the Earth is closest to the Sun is called *perihelion*, and at its farthest is *aphelion*. The Earth and Sun (or any two orbiting bodies) actually orbit around the center of mass of the system. For a system where one object (the Sun, in this case) is so much bigger than the other, the center of mass is actually inside the Sun but not at the center. As the Earth orbits around the Sun, the Sun wobbles a little due to the gravitational pull of the Earth.

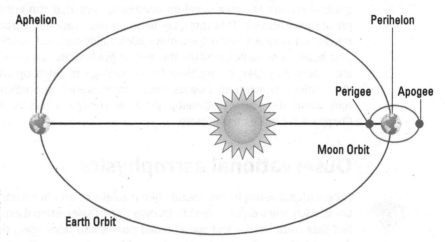

Aphelion   Perihelon

Perigee   Apogee

Moon Orbit

FIGURE 4-3:
Schematic diagram of the Earth's orbit.

Earth Orbit

Similar effects are seen from the Moon's orbit of Earth. The elliptical orbit of the Moon around the Earth is called *perigee* at closest approach, and *apogee* at farthest (see Figure 4-3). The apparent angular size of the Moon in the sky increases about 14 percent at perigee. A full Moon that takes place at perigee is called a supermoon and is about 30 percent brighter than a full Moon at apogee. The gravitational pull from the Moon tugs on the Earth, causing tides. This gravitational tugging is seen best in the oceans, waves of water that rise and fall on a twice daily cycle because water is much more easily deformed than the solid rock of the rest of Earth's surface. See Chapter 6 for more on orbits and gravity.

Diving into the Details of Astrophysics

Astrophysics as a subject area is quite broad. It applies physical laws and principles to bodies observed via astronomy. There've been nearly a billion astronomical bodies noted to date, and that's a vast amount of material to study. Dividing and conquering is often the best strategy when presented with a problem too large to attack all at once, and certain aspects of astrophysics gravitate more toward specific methods of study than others. The following sections take a look at a few of the major subfields to see how their methods and approaches differ.

Theoretical astrophysics

TIP

The subset of astrophysics known as theoretical astrophysics studies objects and phenomena observed via astronomy. Scientists use physical and mathematical modeling to explain these observations. Theorists also make predictions that help guide observers to make novel observations, ones that can either prove or disprove their theories. This interplay between observational astronomy and theoretical astrophysics has helped move both fields forward, particularly over the past hundred or so years. Some theoretical predictions can be tested right away, but others may take a long time for technology to catch up with the required observations. Gravitational waves, for example, were first predicted in 1916, but it took about 100 years to develop sensitive enough detectors. Surf on over to Chapter 8 for more on this story.

Observational astrophysics

TECHNICAL
STUFF

Observational astrophysics sounds like it might be another name for astronomy, but it's actually a distinct field in its own right. Observational astrophysicists collect data from Earth- and space-based probes and telescopes, then analyze that data with the fundamental principles and laws of physics. Think of observational astrophysics as a way of testing and confirming theoretical astrophysics as researchers compare the sky above with the model sky in their computer models.

Laboratory astrophysics

Laboratory astrophysics sounds like anything that can be tested in a laboratory, right? Almost! This field of study works from the scale of particles to the molecular level. Researchers dig into the nature of molecular clouds, planetary composition, light spectra, and more. Laboratory astrophysics is closely related to experimental astrophysics, a field in which scientists conduct experiments and tests on various phenomena in order to validate theories. Tools used in both fields include particle accelerators, lasers, and advanced telescopes.

High-energy astrophysics

TIP

Although many things in space are larger than life, some are metaphorical boulders in a field of pebbles. The branch of astrophysics called high-energy astrophysics uses space- and Earth-based telescopes to study bodies and phenomena that give off the shortest wavelength radiation, such as x-rays and gamma rays. Major targets of study for extreme astrophysics include neutron stars and black holes, as well as cosmic rays, gamma ray bursts, and active galactic nuclei. This field also merges certain techniques from particle physics in laboratory experiments.

Running the gamut of the universe with cosmology

A final subfield of astrophysics is cosmology, the study of universe at the largest scales of time and size. If you're curious about how the universe got its start, and what the Big Bang is all about, cosmology is the field with the answers — and Chapter 13 has a few of them. Cosmology is also concerned with the end of everything. What will happen to the universe in the far-off future? Will it keep expanding forever or contract back down to a point, or might it do something entirely different?

Astrophysics may not have any definitive answers for you on the ultimate fate of the universe, but it does have some intriguing speculations. See Chapter 16 for more on this brain-shattering topic.

The Nitty-Gritty of Telescopes and Astronomical Instruments

How do scientists take all of those observations? Astronomers and astrophysicists use a variety of Earth- and space-based telescopes, as well as other instrumentation, in order to observe the cosmos. The type and location of telescopes depends on what part of the electromagnetic spectrum they are intending to observe — an optical telescope looks very different from a radio telescope, for example, because they look at very different wavelengths.

Optical telescopes

Visible-wavelength observations are probably the most familiar to you. Refracting telescopes use a set of lenses to magnify and focus light, but this setup generally requires a very long tube and is limited in size because gravity will deform lenses.

Most current visible wavelength telescopes use mirrors, either a single huge mirror or an array of smaller mirrors, and are called reflecting telescopes. The mirrors have a concave shape that is slightly curved, so the light from a large area is focused to a point. Most telescopes have a secondary mirror that sends the light to a detector at the focal point. Figure 4-4 illustrates how basic reflecting and refracting telescopes work. In practice, reflecting telescopes have many forms. Some have a tilted mirror that sends light to the side, some reflect light through a hole in the primary mirror, some do bizarre things like send light on a journey to a room full of equipment!

Refracting telescope:

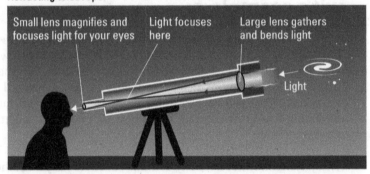

Small lens magnifies and focuses light for your eyes

Light focuses here

Large lens gathers and bends light

Light

Reflecting telescope:

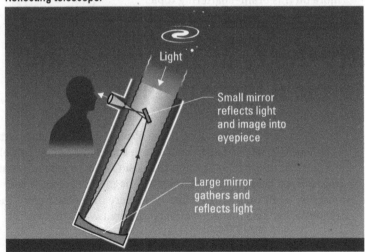

Light

Small mirror reflects light and image into eyepiece

Large mirror gathers and reflects light

FIGURE 4-4:
Diagram of a reflecting and refracting telescope.

One of the advantages of a reflecting telescope is that you don't have the chromatic aberration that you would see with a refracting telescope. All wavelengths of visible light reflect similarly and don't spread apart, as they would by the lenses of a refracting telescope. The mirror of a reflecting telescope must be manufactured to extremely tight shape tolerances; they also need to be polished to create a bright reflecting surface that can capture as much light as possible.

WARNING

Professional telescopes are usually located in large domes that are closed during the day to protect the telescope but open at night to allow the starlight to reach the telescope's mirror. Contrary to popular belief, telescopes don't actually poke out of the door of a dome. A portion of the dome opens over the telescope, and the dome and telescope can rotate in order to see all portions of the night sky. The dome and telescope have motors to rotate and reorient in order to smoothly follow the motion of the sky over the dome, the goal being to track and observe distant astronomical objects over the course of minutes or hours.

In addition to the optics, a telescope can also have a variety of detectors attached to it in order to actually take the observations. Some observers might want to take images in a clear filter or in a number of narrow-band color filters. These can then be combined to create color images and to take color ratios that determine basic properties of the object.

Software allows this data to be used for science in many ways, including

>> **Photometry:** An extremely accurate measurement of the brightness of objects

>> **Astrometry:** An extremely accurate measurement of the locations of objects

TECHNICAL STUFF

Some observers use a spectrometer, an instrument that splits the light from a distant star or galaxy into its brightness over a broad range of wavelengths. This spectral signature can be diagnostic of the chemical composition of the source object.

TIP

Some telescopes are specially designed to search for objects (such as comets or asteroids) moving in the night sky against the background of stars. The search for asteroids is particularly important because it could provide warning of a potentially hazardous object on a collision course for Earth — an important threat to identify sooner rather than later!

The night life at optical observatories

Currently, the largest optical telescope in the world is the Gran Telescopio Canarias. It's located at an observatory on La Palma in the Canary Islands of Spain. This telescope has a single mirror with a diameter of 34.1 feet (10.4 meters). Optical telescopes work best at high altitudes, on mountaintops far away from city lights. High-altitude locations are optimal because the atmosphere is thinner; benefits of higher altitude include the following:

>> Fewer distortions of the observations due to atmospheric motions

>> Less absorption by the atmosphere itself (especially water vapor)

>> A dark-sky location, also essential to avoid light pollution

Other types of telescopes

In addition to optical telescopes, professional astronomers and observational astrophysicists use a number of other types of telescopes. Radio telescopes observe at longer radio wavelengths. These observatories look like large satellite dishes, and their resolving power is based on their diameter, similar to optical telescopes — but because the wavelength of radio waves is much longer than visible light waves, the telescopes must be much larger.

TIP

ADAPTIVE OPTICS

Although a high-altitude location can help reduce fuzziness in a ground-based telescope's images due to atmospheric turbulence, it's impossible to completely remove this factor except by going into space, completely outside the atmosphere (see the next section). However, because space telescopes are very expensive and difficult to update or service, researchers have developed a system called adaptive optics, which allows a telescope to compensate for atmospheric motion in real time. A telescope with adaptive optics (AO) measures the distortions due to the atmosphere in real time, and will actually distort the shape of the telescope's primary or secondary mirror using an array of tiny motors to reposition segments of the mirror. The computer controlling the AO system monitors the seeing (the amount of atmospheric distortion) and corrects for it using a deformable mirror, to produce sharp images rather than ones that are blurred out by distortions. The system requires the use of a nearby bright guide star to measure the distortion, although a laser beam can also be used to generate an artificial guide star.

TECHNICAL STUFF

Radio telescopes can actually be arrayed together using a process called interferometry to produce an effective diameter that is even bigger than that of a single dish. The Very Large Array (VLA) in New Mexico is made up of an array of 27 parabolic dishes, each with a diameter of 82 feet (25 meters). Each dish in the array can slide along huge rails set in the ground to be positioned to change the resolution and effective wavelength. A dedicated control system integrates the signals from all 26 dishes simultaneously, and this results in a signal equivalent to a single radio telescope dish with a diameter of 22 miles (36 km)! Figure 4-5 shows some of the telescopes in the VLA.

FIGURE 4-5: The Very Large Array, a radio telescope array in New Mexico.

Courtesy of NRAO/AUI/NSF

Another telescope array, the Atacama Large Millimeter / submillimeter Array (ALMA), is located in the Atacama Desert of Chile. This array measures signals that are shorter in wavelength than radio signals but longer than visible light. It consists of 50 antennas with diameters of 39 feet (12 meters), plus 16 other telescopes with smaller diameters. These all work together as a giant interferometer. Although interferometry is theoretically possible at any wavelength, it gets harder to do as wavelengths get smaller, and interferometry is rarely done with IR or visible light, and never at shorter wavelengths.

One advantage of radio and millimeter wave telescopes is that they can operate 24 hours a day; they don't have to wait until the sun goes down to take their observations. Although visible light pollution is not an issue at radio wavelengths, radio frequency interference (RFI) is a huge deal.

WARNING

Visitors to radio telescopes must make sure that all electronic devices, including cell phones and digital cameras, are switched completely off while visiting. These are some of the last places you'll find wired phone connections.

Another type of telescope that can operate during the day is a solar telescope. Solar scopes have one main purpose: to observe the Sun. They can be used to monitor sunspots, solar prominences, and other important heliophysics properties (see the later section "The Sun, the Star of Our Solar System" for more on the Sun as a star). Although most astronomical telescopes try to amplify the light from very faint objects, a solar telescope has the opposite problem — it must use special filters to block out much of the Sun's light, so that the telescope doesn't overheat or melt!

TECHNICAL
STUFF

Other types of detectors measure exotic particles such as neutrinos. These rare particles need a huge detector such as IceCube, a particle detector made out of a cubic kilometer of pure ice, located below the surface of Antarctica. Another neutrino detector is the Super-Kamiokande observatory, in Japan, which consists of 50,000 tons of pure water buried in a structure deep underground. The ice or water detectors in both cases are surrounded by an array of very sensitive light detectors, which look for the faint flashes of light given off when a lone neutrino interacts with the ice or water. The IceCube telescope detects about 275 neutrinos a day, and these systems can also detect cosmic rays.

Viewing from above: Space-based telescopes

Although ground-based telescopes have many advantages, there are also reasons to operate telescopes in space. Space-based optical telescopes like the venerable Hubble Space Telescope have a number of advantages. They can observe 24 hours a day and don't have to wait for nighttime for observations. Telescopes in space also don't have that pesky atmosphere to look through, meaning that their observing power is constrained only by the diffraction limit.

The diffraction limit is the theoretical maximum for the angular resolution of a telescope and is defined as the wavelength of light divided by the diameter of the mirror. In the case of Hubble, its 2.4-meter mirror means that it can distinguish two stars that are only 0.05 arcseconds apart at visible wavelengths. This capability is 20 times better than what we can generally see from the ground. To do better, you need systems like the four-telescope Very Large Telescope, which has a resolution of 0.002 arcseconds . . . but is also four 8-meter mirrors working together and using adaptive optics!

TECHNICAL STUFF

Space-based telescopes can also observe at wavelengths that are not visible from the ground due to absorptions by the Earth's atmosphere. Our atmosphere absorbs much of the harmful high-energy radiation produced by the Sun and by cosmic rays, and this absorption is good for the survival of life on Earth. It can, however, be inconvenient for astronomers! Ultraviolet observations are often made using high-altitude balloons, but space telescopes can give more consistent views. The Hubble Space Telescope can observe in UV wavelengths as well as visible wavelengths.

Space telescopes also operate at shorter wavelengths, such as the Chandra X-ray Observatory, the Compton Gamma Ray Observatory, and the Fermi Gamma-Ray Space Telescope. Longer-wavelength space telescopes include the ESA Herschel Space Observatory and the Spitzer Space Telescope, both of which observe at infrared wavelengths.

The newest infrared space telescope is the James Webb Space Telescope, which began operations in 2022 and is larger and more sensitive than any previous space telescope. With its enhanced observing power, JWST is making key discoveries in many aspects of astronomy and astrophysics, many of which will be discussed in later chapters.

The Sun, the Star of Our Solar System

Before we delve into the field of astrophysics as a whole, we first start with our very own star, the Sun. Held together by gravity and burning over 27 million degrees at its core, the Sun is one hot potato. Radiating that much solar energy gives the sun a rather magnetic personality — literally! The field of heliophysics is devoted to studying the internal workings of the Sun, as well as the mechanisms by which the Sun interacts with its surrounding planets and atmospheres.

Why study heliophysics, you ask? Space weather results from charged particles and radiation emitted by the Sun interacting with other things, and its dynamic changes affect the Earth. It's also a great idea to use what you have; because we're able to study the Sun in so much depth, scientists take what they learn from the Sun's interaction with other planets and apply it elsewhere in our galaxy and beyond.

The Sun is, arguably, one of most important influencers out there. Far more interesting than providing opinions on restaurants or shoes, the Sun influences our Moon and Earth with its modulating magnetic fields and solar winds; the science of heliophysics is devoted to those details of the Sun's properties. We use an array of spacecraft — the Parker Solar Probe, for example — to provide us with data that helps us better understand the effects of the Sun, and, in particular, how it interacts with the Earth.

Solar flares, solar wind, and other solar activity

Our Sun is constantly emitting particles such as electrons and protons from its corona (the outermost part of its atmosphere). These particle outpourings are collectively known as the solar wind. They are magnetically charged thanks to the magnetic fields emanating from our Sun, and they are considerably faster than your regular Earth wind — the solar wind travels from 250–500 miles per second. As a constantly-rotating star, the Sun emits these streams of solar wind as a radiating spiral arm, and they reach Earth via our world's lines of magnetic field.

Magnets can be charged positively or negatively. The negative pole on one magnet attracts the positive pole on another, hence the saying "opposites attract." When talking about the Sun's magnetic fields, you're looking at attraction on steroids. The Earth has its own internal magnetism that creates a magnetic field surrounding our planet, the magnetosphere. The Sun also creates powerful magnetic fields that flow outward into space, and they interact with the Earth's magnetosphere.

Nothing can throw off a solar wind like a good solar flare! Imagine a toddler stomping her foot in an outburst. Now, imagine that child producing 10^{25} joules worth of energy with every stomp, and you've got yourself a solar flare (10^{25} is 10,000,000,000,000,000,000,000,000 joules — that's more energy than used by modern humans!). These flares occur in active areas of the Sun, including locations where the Sun's magnetic field is enhanced such as in sunspots, and shoot off both electromagnetic radiation and particles. Due to their magnetic nature, these flares can affect both weather and satellite communications here on Earth. Figure 4-6 shows a mid-level solar flare, as seen by the Solar Dynamics Observatory spacecraft.

While you're at it, be on the lookout for coronal mass ejections (CME). These emissions are similar to solar flares but on a large scale — millions to billions of tons of particulate material, some carrying magnetic fields, can be fired out from the Sun's corona (hence the name) during a CME event. Although they can be damaging to satellites, when a CME arrives at Earth, it can also produce beautiful auroras in the northern and southern latitudes.

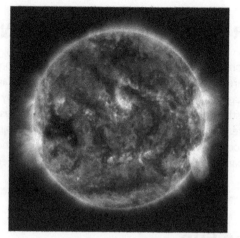

TIP

Want to observe a coronal mass ejection? A coronagraph is a special instrument that blocks out enough of the Sun's light that we can actually study its corona. The Keck telescope in Hawaii, for example, has such an instrument today, and more are planned for future space telescopes. You can also safely see the Sun's corona during a total solar eclipse — more on that in a moment!

Hot or cold? The ins and outs of weather in space

Space is a vacuum, right? As you see in Chapter 2, vacuums are created in the absence of matter. Here on Earth, we are literally full of air and water — very much not a vacuum. Earth-based weather relies heavily on understanding how our atmosphere plays with air pressure, temperature, and other conditions that create storms (perfect or otherwise) and more. Space-based weather, though, doesn't concern itself with those nitpicky requirements for human life, but it still has its stormy days

Weather in space is based around the charged particular matter bursting from the sun via solar wind and flares. A good old-fashioned space weather storm kicks up when flares or CMEs are too large to be deflected by our magnetosphere and, instead, crash right into it. Charged particles rain down across our magnetic field lines and interact with our atmosphere in potentially brilliant ways. The auroras visible from the Earth's poles? Yep — those are examples of solar wind particles making friends with our atmosphere.

HOW SOLAR FLARES AFFECT LIFE ON EARTH

As solar wind travels from the Sun outward, greater or fewer particles may reach us on Earth. During a solar flare or CME, solar wind may pick up the pace and arrive on Earth with more force than usual. If you're a fan of weather prediction, you rely on satellites to provide us Earthlings with forewarnings. What happens when solar flares interact with those satellites? You guessed it: interruptions to our satellite communication system. Solar flares can also knock out our precarious Earth-based power grids because those magnetically-charged particles from the Sun can cause failure of power equipment and radio communications.

Studying the Sun

You might think that it's easy to observe the Sun — just look up in the sky, right? Some observations can only be made from close proximity. One of the first spacecraft solar observation missions was the ESA Solar and Heliospheric Observatory (SOHO), launched in 1995 and still going strong today. SOHO studies the Sun's outer layers (including the chromosphere and corona) using a variety of remote sensing instruments. It also observes the solar wind, and studies the interior structure of the Sun using helioseismology. SOHO is unusual because it's actually located near the Sun-Earth Lagrange point L1, a stable location between the Earth and the Sun where the Sun's and Earth's gravity balance out. Coincidentally, SOHO has also discovered around 5000 comets, most of which are on a path to destruction as they approach the Sun.

Another early solar observation mission was Ulysses, a joint ESA/NASA mission that actually orbited the Sun for an increased range of views. Ulysses launched in 1990 and operated until 2009. In order to reach its high inclination orbit of 80 degrees for viewing the Sun's poles, it had to use a long gravity-assist trajectory by looping out around Jupiter before heading back towards the Sun. Data from Ulysses provided important insights into the magnetic field of the Sun, as well as offering unique observations of gamma rays and cosmic dust originating outside the solar system.

TIP

One modern solar-observing spacecraft is the Parker Solar Probe, launched in 2018 and still operating today. Parker used multiple flybys of Venus to reach an orbit that allowed it to study the low parts of the solar corona, the Sun's outer atmosphere. It's currently studying

» The plasma and magnetic field of the corona

» How heating of the corona triggers the solar wind

» Ways in which energetic particles of radiation are accelerated by processes initiated in the corona

Eclipses, or Throwing Shade in a Scientific Way

Eclipses are one of the most alluring astronomical phenomena out there. They occur rarely and predictably enough that you've got time to prepare and get to an optimal location for viewing, and they're so visible that they're almost impossible to miss. They're also a stark, visible reminder of the incalculably vast nature of space.

TECHNICAL STUFF

We are but a pebble in the cosmos; the Moon's diameter is 30 times that of Earth, and the Sun's is over 100 times as large. The enormity of an eclipse makes you realize just how tiny we are with respect to the passage of these bodies around us.

What exactly is an eclipse? All eclipses occur when the light from one object is blocked out by another. A solar eclipse takes place when the Moon comes between the Earth and the Sun, and a lunar eclipse results from the Moon moving into the Earth's shadow as it's blocked from sunlight.

Red Moon: Lunar eclipses

REMEMBER

The Earth orbits our Sun every 365 days, and rotates about its own axis every 24 hours. At the same time, the Moon orbits the Earth in a 27.5 day orbit relative to the Sun, though because the Earth is always in motion around the Sun, the Moon actually takes 29.5 days to complete its orbit relative to the stars. At certain points during these orbits, the Earth occasionally moves in between the Sun and Moon, effectively preventing Sunlight from reaching the Moon.

TECHNICAL STUFF

Don't think of the Moon as a massive glow-in-the-dark ball. The Moon appears illuminated because it reflects light from the Sun. When the Sun's light is prevented from illuminating the surface of the Moon, the Earth casts a shadow onto the Moon. This phenomenon is what scientists call a lunar eclipse, and it's illustrated in Figure 4-7.

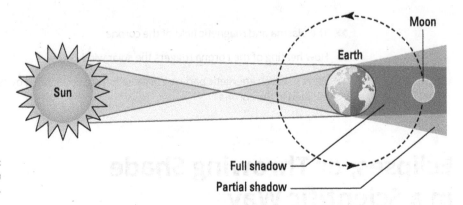

FIGURE 4-7:
Diagram of a
lunar eclipse.

Labels in figure: Moon, Earth, Sun, Full shadow, Partial shadow

There are three basic classes of lunar eclipses:

>> **Partial lunar eclipse:** Occurs when the Moon orbits into the Earth's shadow, but not completely.

>> **Penumbral lunar eclipse:** Occurs only when the Moon is passing through the penumbra (outside shadow) of the Earth.

>> **Total lunar eclipse:** Occurs rarely, and requires the Sun and Moon to be precisely on opposite sides of our planet. Sunlight going into the atmosphere surrounding Earth is filtered, removing blue light and leaving a reddish haze of light to reflect off the Moon during the eclipse.

A lunar eclipse always happens during the full Moon because this is the time when the Sun, Moon, and Earth are aligned. Lunar eclipses are rare, but not that rare; on average, for a given location, a total lunar eclipse happens about every two or three years. Lunar eclipses occur at night because that's when the full Moon will be visible. Over the course of several hours, you see a dark shadow start to move across the face of the bright Moon. You don't need any special glasses or equipment to observe lunar eclipses; your unaided eyeballs are good enough, although binoculars or a small telescope will give a sharper view.

When the Moon is completely covered by the Earth's shadow, you've reached the phase known as *totality*. In a partial eclipse, only part of the Moon will be covered. If it's a total eclipse, the entire Moon will be in shadow. Its appearance will depend on atmospheric conditions, but sometimes the whole Moon will have an eerie red or yellow appearance. Slowly, the Moon will re-emerge from the Earth's shadow and become brighter and brighter until eventually it's fully illuminated again. Figure 4-9 shows a collage of images taken during a lunar eclipse to illustrate the different stages.

Don't look! Solar eclipses

A lunar eclipse is an unusual sight but a solar eclipse, particularly a total solar eclipse, can be a once-in-a-lifetime event. Solar eclipses occur when the Moon passes between the Sun and Earth, blocking the Sun's rays from reaching us on Earth. Figure 4-8 shows the geometry of a solar eclipse. These are the main types:

>> **Total:** Direct alignment of Earth, Moon and Sun, also called a "blackout eclipse" when the Sun is completely covered for a few minutes and you are in a shadow.

>> **Annular:** The Moon appears as a dark circle over the Sun because it's too far away from the Earth to completely block the sun.

>> **Partial:** The Moon blocks part of the Sun's disk.

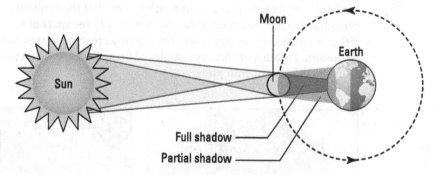

FIGURE 4-8:
Diagram of a solar eclipse.

Although solar eclipses are rare, a partial eclipse is more common than an annular or total eclipse. In a partial eclipse the Moon passes in front of the Sun but only partially obscures it because it passes either too high or too low to fully cover it. An annular eclipse occurs when the Moon is farther away from the Earth in its orbital cycle, appears smaller, and can't fully cover the Sun; these look like a bright ring with a dark circle in the middle. Finally, a total eclipse happens when the Sun is completely obscured. It's possible for one event to have all three kinds of eclipses! People farther from the center of the Moon's shadow on the Earth will see a partial eclipse, and sometimes the Moon moves far enough away during an eclipse that some people will see an annular at the start or end of the event, whereas others only see a total eclipse.

WARNING

Observing a solar eclipse of any kind — partial, annular, or total — requires special equipment to protect your eyesight. You need to use eclipse-certified viewing glasses. Unless you're absolutely certain that your glasses are certified for use during an eclipse, don't use them (blindness is no joke!). If you have a telescope or binoculars, you need a special solar filter called a neutral density filter or Hydrogen filter for your telescope in order to observe the eclipse directly.

If you don't have appropriate eye protection or telescope protection, however, there are still safe ways to watch an eclipse. You can use your telescope or binoculars to project an image of the Sun onto a floor, wall, or screen in order to observe an eclipse's progress. You can also make a pinhole viewer by using your fingers to create tiny holes in a card and look at your hand's shadow. This even works with a spaghetti strainer!

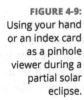

TIP

A fun activity with kids is to draw a shape or a letter on a card, and then punch a pattern of pinholes in the card following the drawing. During the eclipse, hold the card out with a second card underneath it such that the sunlight goes through the pinholes. Try this on a normal day and you'll only see spots of light; during a partial or annular eclipse, you'll actually see tiny crescent shapes that are the shape of the Sun! Figure 4-9 shows examples of using your hand as a pinhole viewer, or using an index card with pinholes.

FIGURE 4-9:
Using your hand or an index card as a pinhole viewer during a partial solar eclipse.

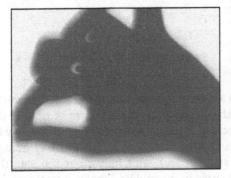 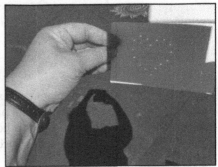

Courtesy of Shana Priwer and Cynthia Phillips

During an annular eclipse, at its maximum coverage the Moon's disk will be completely in front of the Sun but won't block it out completely. What's left is a narrow bright circular ring around the Sun's outside. This ring is still bright enough to harm your eyes, so don't remove your eye protection! Figure 4-10 (left) shows a view of an annular eclipse.

FIGURE 4-10:
Annular and total
solar eclipses.

Courtesy of Stefan Seip (left) and NASA / Aubrey Gemignani (right)

If you are lucky enough to be observing a total solar eclipse, you'll see the sky getting darker and darker as the eclipse approaches totality. When the Moon is completely covering the Sun, it's safe to remove your eye protection. You may observe that the temperature momentarily drops; birds and other animals may act strangely because they'll think that it's night!

REMEMBER

You may be able to see the corona during totality. Recall that the Sun's corona is its outer extended layers that are usually obscured by the Sun's bright light. Refer to Figure 4-10 (right) for an idea of what that might look like.

Totality only lasts a few minutes. Keep an eye on the time and put your eye protection back on before the Sun is scheduled to emerge. As the Moon's shadow moves off the Sun, you may see what's called a diamond ring effect when the first rays of sunlight emerge between lunar mountains.

ECLIPSE TIMING

Solar eclipses occur about every 18 months somewhere on Earth, and are fast — often just 2 to 4 minutes will pass before totality is over. Lunar eclipses, on the other hand, occur more often (about twice a year for partial lunar eclipses) and last 30 minutes to an hour, so there's more time to get outside and start observing.

Science of eclipses

In addition to being an awe-inspiring way to connect with the heavens, eclipses bring great opportunities for science. Lunar eclipses were used by the ancient Greeks to determine that the Earth was spherical, and Aristarchus estimated the diameter of the Moon using the timing of lunar eclipses in the 3rd century BCE. Today, lunar eclipses help with studying how the atmosphere refracts different colored light in different conditions (a lot of smoke or volcanic ash makes a redder Moon!).

Solar eclipses, not to be outdone, are also opportunities for amazing science. For example, although you can use a coronagraph to view the outer layers of the Sun's atmosphere, you can't see as much as scientists would like. During an eclipse is the only time astronomers can observe the inner part of the corona, the originating source of solar wind particles that cause space weather here on Earth. Researchers can also study the ozone layer's location and thickness by looking at how the atmosphere responds to being heated and cooled by the Moon's passing shadow.

TIP

Spectroscopic observations of the Sun's corona during eclipses in the mid-1800s were used to discover the presence of hydrogen and helium in the Sun's outermost atmosphere. In 1919, British astrophysicist Arthur Eddington used observations of a solar eclipse to confirm Einstein's theory of general relativity. Eddington was able to show how space-time is distorted in a gravity field by making careful observations of stars near the surface of the Sun during the eclipse, and comparing them to images taken when the Sun was not in that part of the sky. This showed how the Sun's gravity can bend light and change where stars appear. These observations matched predictions made by Einstein. Curious for more? Head over to Chapter 15 for details on relativity.

Astrophysicists today study the Sun's inner corona to better model heat and energy transfer within the Sun. In turn, these models improve our understanding of how solar wind is generated by providing detailed models of its source region. Corona models are also used to help predict the impact of space weather on Earth.

TIP

During a solar eclipse, scientists can observe how the Sun affects the Earth's upper atmosphere by monitoring its variation as the illumination changes quickly. Energy from the Sun impacting the Earth's atmosphere creates the ionosphere, an outer layer of charged particles. Because many satellites in low Earth orbit are located within the ionosphere, understanding its generation and dynamics is important for satellites safe operation. Other measurements made during eclipses can help verify models of the creation of the thermosphere, another outer layer of the Earth's atmosphere. All these measurements help yield insights into our very own star, the Sun, and provide precious information that anchors the field of astrophysics.

Now that you've gotten a (sky) high-level overview of how the fields of physics and astronomy come together to produce astrophysics, it's time to dig into the details. First up will be a look at other stars, both like and unlike our Sun.

2

When You Wish Upon a . . .

Learn how stars are born and how you can tell them apart, and understand different scenarios for their ultimate demise.

Gain knowledge of the dynamics of multiple star systems, how star clusters work, and why "interstellar gas and dust" is the most often-used phrase in this book.

Discover how exoplanets, planets beyond our own solar system, are found, and explore how we're searching for Earth 2.0.

See why the galaxy is filled with weird objects such as white dwarfs, black holes, gravitational waves, quasars, and more.

Chapter 5

Star Power: Hydrogen, Helium with a Twist of Nuclear Fusion

I f you've ever looked up into the night sky and wondered what makes those stars shine, you've come to the right place. Understanding the heavenly bodies in our cosmos might present more questions than answers but, as they say, stars tell a tale as old as time. Stars provide the light and heat (Hello, Sun!) that make life on Earth possible. Without stars, there would be no people. In addition to the all-important role of — well, allowing Earth to be habitable — stars provided the earliest human navigators with a means of finding their way. They offer comfort and familiarity to all who view them, and they are the basis for countless cultural traditions around the world.

How did the first star come to appear in the sky? The birth of the stars opens a window to the history of the universe. Physically and scientifically, stars represent the complexity that comprises our galaxy. Understanding the evolution of both single stars and ones clustered into galaxies provides a basic understanding of the cosmos. As with all things, stars don't live forever. Even in death, though, stars show us the forces and reactions that are the very fabric of our universe, and create the building blocks for life itself. From the infancy of a protostar to its rather brilliant ending in a supernova, stars tell a story — the ultimate story of astronomy and physics, to be sure!

Happy Birthday! A Star Is Born

Perhaps the most-observed objects in the night sky, stars are not few and far between. There are more than 100 billion stars in our Milky Way Galaxy. Supposing that there are about this many in any typical galaxy, and given that there are at least 200 billion galaxies in the visible universe, how many stars are there in our sky? Upwards of 200 sextillion (or billion trillion; in other words, a 1 with 21 zeros after it). Far too many to count, that's for sure. But what exactly constitutes a star?

How do we define stars?

Contrary to your childhood drawings, stars aren't symmetric two-dimensional objects with sharp points. In real life, stars are actually luminous balls of (mostly) hydrogen and helium gas.

So why does a star shine? The answer lies in fusion, and not the sort that you create with delicious multicultural meals — we're talking about nuclear fusion. Stars, like our Sun, are so massive that they are compacted by their own gravity. At the core of a star, the temperature and pressure rise to insanely high levels. At the center of our Sun (which is considered only a mid-sized star by galactic standards), the temperature is about 27 million degrees F (15 million degrees C) and the pressure is 100 billion atmospheres. This much temperature and pressure inside the Sun's core prompts thermonuclear fusion.

TECHNICAL STUFF

You're probably familiar with the term *atmosphere* as the protective, multi-layered structure of gases that surrounds Earth and other planets, but an *atmosphere* is also a unit of measure. In this context, one atmosphere (atm) is defined as the air pressure at sea level; its value is 14.7 psi (pounds per square inch) or 101,325 Pa (pascals).

Two is better than one: Fusion

Thermonuclear fusion is a process that breaks all the normal rules of chemistry *because it is physics!* Let's take hydrogen as an example. At the crazy-high temperatures and pressures found inside a star, two hydrogen atoms can actually fuse together to make an atom of a second, heavier element. Hydrogen is the simplest and most common element in the universe, making it the most common fuel for stars — deep inside a star, two hydrogen atoms can fuse together to make one helium atom when conditions are right. This process releases huge amounts of energy because the mass of the single nucleus at the end of the fusion process is less than the masses of the two initial atoms. This extra mass is released as photons that carry away energy, following Einstein's famous formula $E=mc^2$.

REMEMBER

Einstein's formula for mass-energy equivalence, $E=mc^2$, shows the relationship between mass (m) and energy (E). If a little mass is converted into energy, it yields a lot of energy because the c in this equation is the speed of light — a huge number! The kind of nuclear fusion that takes place deep inside our Sun is called proton–proton fusion because although it involves hydrogen nuclei, a hydrogen nucleus is nothing more than a proton. These are the main steps in proton–proton fusion (see Figure 5-1):

1. **Two protons fuse together, and the weak nuclear force causes one of the protons to turn into a neutron.**

   This process releases a positron (basically an electron with a positive charge) and a neutrino (a tiny, non-interacting particle with no charge). The resulting neutron-proton pair is called *deuterium*. Deuterium is an isotope, or different kind, of hydrogen — normal hydrogen just has one proton and one electron, whereas deuterium also has a neutron.

2. **Another proton collides with the deuterium, creating what's called a helium-3 nucleus.**

   Helium-3 is a lighter isotope of helium that has two protons but only one neutron. This collision releases a gamma-ray photon.

3. **Two of these helium-3 nuclei crash into each other, and this collision produces a helium-4 nucleus plus two extra protons.**

   Helium-4 is the more common kind of helium, which contains two protons and two neutrons in its nucleus. The extra protons are released as two hydrogen nuclei, which can then start the process all over again.

The gamma rays released in Step 2 are some of the energy that travels through the Sun out into space. Assuming you like the fine art of existing, thank this energy — it supports all life on Earth as sunlight. The helium-4 atom that's formed in Step 3 was made up of two protons and two neutrons, but it had less mass than the original particles. This extra mass is also released in the form of energy, specifically heat and light. It travels from the center of the Sun through the Sun's many layers and eventually out through space to the Earth.

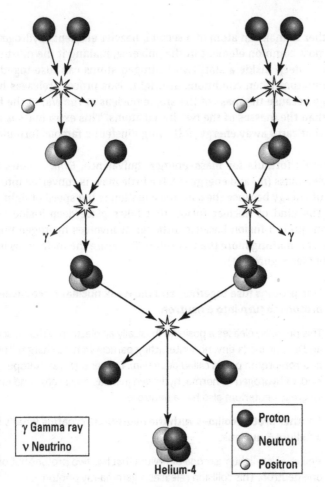

FIGURE 5-1:
Solar nuclear
fusion diagram —
from hydrogen
to helium.

| γ Gamma ray |
| ν Neutrino |

| ● Proton |
| ● Neutron |
| ○ Positron |

Helium-4

The proton–proton reaction in which hydrogen is fused into helium is not a rare event. It takes place 9.2×10^{37} times per second, meaning a huge amount of energy is released by the Sun every second. Even though only about 0.7 percent of the mass in the fusion process ends up becoming energy, the Sun is so big that 4.7 million tons (4.26 million metric tons) of mass convert to energy every second, resulting in a release of about 5×10^{23} horsepower or 3.8×10^{26} joules of energy per second. The process of creating heavier elements out of lighter ones through nuclear fusion is called *nucleosynthesis*, and it's how all the heavier elements in our universe were created.

REMEMBER

All these steps can take place only at the high temperatures and pressures found deep inside a star, or in an uncontrolled way via a thermonuclear explosion. Scientists have been trying to figure out how to make fusion happen in a controlled way in a laboratory, to create a rather large, safe energy source. This topic has been researched extensively, though without success so far.

Protostars, star formation regions, and accretion disks

Lights . . . camera . . . action . . . . Dust cloud? The birth of a star begins with the gravitational collapse of a cloud of gas and dust, far off in the depths of space. Although stars evolve all over the universe, the areas where you're more likely to see stellar objects develop are known as *star formation regions* — Figure 5-2 shows an iconic view of a star formation region in the Carina Nebula, as seen by the JWST. The embryonic phase of a star's initial development takes place in one of these regions. We call those initial clouds of dust and gas from which stars emerged *molecular clouds* because most of the hydrogen there is in the form of the H_2 molecule.

FIGURE 5-2:
Star formation region in the Carina Nebula, as viewed by the James Webb Space Telescope.

Courtesy of NASA, ESA, CSA, and STScI

TIP

A molecular cloud will stay in hydrostatic equilibrium, neither contracting nor expanding, as long as there is a balance between the gravitational potential energy pulling the cloud inward and the kinetic energy of the gas pressure pushing the gas cloud outward.

After the gas cloud gets so large that the gravitational force overwhelms its gas pressure, the cloud will start to collapse. The mass threshold for this collapse is known as the *Jeans mass,* after British astrophysicist James Jeans, and it's related to the temperature, density, and composition of the gas cloud.

TECHNICAL STUFF

The formula for the Jeans mass M_J is

$$M_J = \frac{\sqrt[5]{\frac{15}{\pi}} \left(\frac{k/G}{\mu} \right)^{3/2} T^{3/2}}{\sqrt[2]{\rho}}$$

In this formula, k is the Boltzmann constant (which relates the average thermal energy of particles in a gas to the temperature), G is the gravitational constant (which relates the gravitational force between particles to their mass and distance), T is the temperature, μ is the mean mass per particle, and ρ is the mass density. This formula shows that the Jeans mass M_J is directly proportional to the temperature, and is inversely proportional to the density.

After collapse takes place, gas and dust gather and contract under their own gravitational force into a gravitationally bound core. Molecular clouds are usually thousands to tens of thousands times the mass of our Sun. During collapse, thousands of stars can form simultaneously. The star formation processes can also be kicked off by the collision of two or more molecular clouds, a galaxy collision, a supernova in the vicinity, or even emissions from a black hole at the center of a galaxy. Bottom line: Something shocks a stable molecular cloud into collapsing.

TIP

In addition to regular molecular clouds, there also are thought to be more than 6,000 giant molecular clouds, each containing more than 100,000 times the mass of our Sun, in the Milky Way Galaxy. These are mostly located near the center of our galaxy.

You might be wondering how gas and dust magically turn into stars. As usual, physics comes to the rescue. As the cloud collapses, it also begins to rotate and form a disk. Over time, material interacting in this disk heats up as particles collide, and it also spirals inward faster and faster! In the center, material piles up — more and more — at the center of this spiral. When there is enough material piled up, the gravitational attraction of this mass makes it get denser and hotter until finally a star ignites.

Looking at this in detail:

>> As material accumulates around the contracting center of the protostar, it forms a rotating disk called an *accretion disk* (see Figure 5-3).

>> Conservation of energy (discussed in more depth in Chapter 2) tells us that the speed of rotation increases toward the center of the new solar system.

>> This speed combines with the centripetal force (a force that describes how an object traveling on a circular path stays on that path) to spread the accretion disk wider and flatter. (Like pizza dough getting tossed in the air.)

>> These disks may last only (only!) a few million years, but they emit infrared and sub-millimeter radiation that allows us to track and study them. (They also flare up in x-ray light.)

>> The "protostar" formative phase of a star's life takes a mere 100,000 to a million years to complete before the star turns on.

>> The protostar phase eventually ends when the star finally reaches a high enough temperature and pressure at its core to begin hydrogen fusion.

>> Finally, the remnants of the accretion disk clump together to form planets or get blown away by the star's light pressure (yes, light is pushing things around — even you — just gently).

Figure 5-3 shows a protostar forming within the dark molecular cloud L1527, as seen in an infrared observation from the JWST, located at the center of the hourglass shape. The accretion disk is seen edge-on as the dark band across the center. Outflows from the newly forming star have cleared out regions from the dark gas cloud, which shine brightly above and below the center. This protostar is probably less than 100,000 years old and hasn't gathered enough mass to turn on nuclear fusion yet — it's likely only 20 to 40 percent of the mass of our Sun so far.

Welcome to the world, baby star!

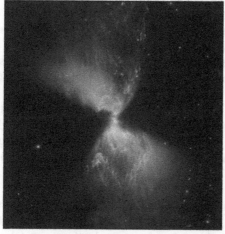

FIGURE 5-3: A protostar forms within the dark molecular cloud L1527, as seen by the James Webb Space Telescope.

Courtesy of NASA, ESA, CSA, and STScI. Image processing: J. DePasquale, A. Pagan, and A. Koekemoer (STScI)

WHY ISN'T JUPITER A STAR?

It's all a question of mass. To be considered a star, celestial objects need enough mass to allow nuclear fusion to begin. Even though Jupiter is made primarily of hydrogen and helium, just like the Sun, it would have needed to be about 80 times more massive in order to have high enough temperatures and pressures at its center to allow hydrogen to start fusing to helium. Jupiter is a very big planet, to be sure, but it's not a star!

Getting to Know Your Stars: Properties, Types, and Characteristics

Astrophysicist Carl Sagan once said that humanity was a way that the universe could understand itself. Demystifying the cosmos, as Sagan so elegantly points out, lets you make the all-important connection between yourself and the universe. From carbon and iron to nitrogen and hydrogen, the fundamental properties of stellar objects can be understood through you — just on a much larger scale!

Don't rely on wishing: Understand the brightness, mass, and more about stars

Stars in the sky are as varied as stars here on Earth. Although the general story of star formation is similar for all stars, the path they take after their formation depends on their composition, location, and most importantly, their mass. Using telescopes here on Earth as well as in space, astronomers have studied stars to determine their observable characteristics and have used these measurements to come up with a series of stellar classes. These classes are a shorthand way of describing a star's current luminosity and color, and even its future evolution.

Table 5-1 shows some of the properties of stars and our Sun that can be measured with a telescope. Astronomers have measured the position, brightness, and other properties of millions of stars — one of the largest star catalogs, which is being created by the European Space Agency's Gaia space telescope, currently has over a billion stars!

If you look up at the stars in the sky, most may appear bright white. Squint (or, better, use a telescope), and you can see that there are some visible differences in color — in the constellation Orion, for example, the star Betelgeuse has a reddish appearance, whereas the nearby star Rigel is noticeably blue.

TIP

You can use color to estimate the temperature of a star. Astronomers do this very accurately using spectroscopy, which comes in handy because you can't exactly bring the stars into your lab. Spectroscopy studies light emission and absorption as it interacts with matter. Specific atoms will act differently at different temperatures, allowing us to accurately measure a star's temperature.

TABLE 5-1 ## Properties of Stars and the Sun

Property	Description	Depends On	Solar Values
Luminosity	The amount of energy a star or other astronomical object radiates, measured per second	Size/mass and temperature	3.8×10^{26} watts (3.8×10^{33} ergs/sec)
Temperature	The surface temperature measured in degrees	Mass	9940°F (5780 K)
Brightness	Apparent magnitude: the brightness of a star as it appears from Earth. Absolute magnitude: the brightness of a star at 10 parsecs (32.6 light-years) away	Distance from Earth and luminosity; temperature	−26.74 apparent magnitude; 4.83 absolute magnitude
Mass	Amount of matter contained in a star	Density, lifetime, properties in related binary star systems	4.4×10^{30} pounds (1.99×10^{33} 99 $\times 10^{33}$ grams)
Size	Usually defined as radius; can be measured directly via imaging for closestr stars	Mass, age	Radius: 2.12×10^{12} feet (6.96×10^{10} cm)
Color / spectral classification	The visible color emitted as a function of the star's temperature; multiple wavelengths	Temperature, composition, and metal content	Yellow; Class G; type G2V
Composition	Chemical elements that make up a star. Primarily hydrogen and helium with other trace components formed in earlier stellar generations.	What kinds of stars came before Age / when formed in rare cases	Hydrogen (91% by number of atoms), helium (8.9% by number), plus traces of nitrogen, oxygen, carbon, and others

This behavior, broadly speaking, defines a star's spectral type. The light from a star is broken down into separate wavelengths using either a prism or an instrument called a diffraction grating to produce a light spectrum. A stellar spectrum (see the image in the color insert section) looks like a spread-out rainbow with dark lines at different wavelengths. These lines are associated with particular chemical elements in the outer atmosphere of the star (the *photosphere*). The appearance of these lines depends on the temperature of the photosphere, and the size of the lines relates to the abundances of different chemical elements.

Astronomers have created catalogs of stars based on their spectral type, as well as on other properties. In the early 1900s, Harvard astronomer Annie Jump Cannon was one of the first to sort stars (350,000 of them!) into categories based on their spectra. Cannon used letters to represent each different class of stars in her category scheme, where stars were grouped using the lines in their atmosphere. Originally in alphabetical order, the letters have been reorganized a few times, but the groupings are still widely used. The category system we rely on today sorts stars using the letters O, B, A, F, G, K, and M — where O-type stars are the hottest and M-type stars are the coolest. This order based on temperature was first established by astronomer Cecilia Payne, also at Harvard, in 1925. Astronomers further subdivide these classes into subclasses by using numbers for the temperature range within a particular type (0 is the hottest, 9 is the coolest), and by using a luminosity class that reflects the surface gravity of the star and whether it is a giant or a dwarf star. We go into more detail on these types of stars in Chapter 8.

Using this classification system, our Sun is type G2V — just an average star on paper, but it means the world to us (literally!)

Plotting magnitude with the Hertzsprung-Russell diagram

Also called an H-R diagram, the Hertzsprung-Russell diagram is one of the most useful ways to see how a star's magnitude (think luminosity and brightness) relates to its spectral classification (think color and temperature). This chart is a combination of plots created by two separate scientists: Enjar Hertzsprung (1911), from Denmark, and Henry Norris Russell (1913), from Princeton in New Jersey. Both came up with this method of visually combining star classification, temperature, luminosity, and magnitude in a single diagram that has effective temperature along the horizontal axis and luminosity along the vertical axis.

It turns out that an H-R diagram is useful not just for comparing the luminosity and temperature of stars, but also for looking at the life-cycles of different types of stars. At the beginning of its life when a star first compacts enough that hydrogen fusion can begin, it has a particular mass. If you put dots on an H-R diagram for the luminosity and temperature of stars from masses ranging from about 60 times the mass of our Sun down to about 0.1 times the mass of our Sun, they form a diagonal line from top-left to bottom-right (see Figure 5-4).

This diagonal line in the H-R diagram is called the *Main Sequence*.

Stars on the Main Sequence are in the prime of their lives. They go merrily about their business, mostly burning hydrogen into helium, but high-mass stars expend energy at a much higher rate than lower-mass stars.

TECHNICAL STUFF

Stars are supported through a careful balance between gravity pushing in and light pushing out. To prevent the masses of the largest stars from collapsing into black holes, stars have to produce a lot of light. When balanced, the luminosity of a star is approximately proportional to mass to the 4th power, or $L \sim M^4$. To maintain this balance, high-mass stars burn their hydrogen fuel at a much faster rate than low-mass stars. Even though they start out bigger to begin with, a high mass star's lifetime as a Main Sequence star (which ends when they run out of hydrogen fuel) is much shorter than that of an average-sized star. In fact, a star with a mass of ten times the mass of our Sun will have a lifetime of about 32 million years, whereas our Sun will have a lifetime of about 10 billion years. It's a good thing for us that our Sun wasn't too massive to begin with! Smaller stars tend to live a lot longer; a star with an initial mass of about 0.1 times the mass of our Sun will last for thousands of billions of years.

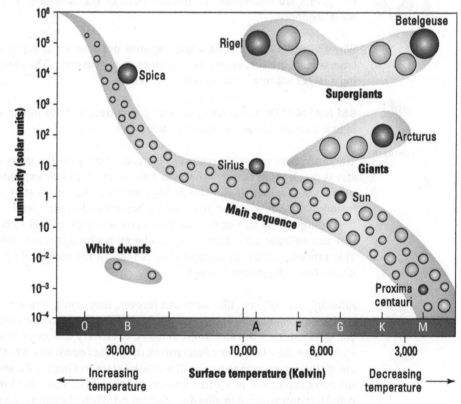

Dwarfs and giants

As stars near the end of their lifetimes, they come to a fork in the road: What will they become next? The answer depends mostly on a star's initial mass. For a typical yellow dwarf star like our Sun, after it has used up all the hydrogen fuel in its core, the nuclear reactions that have been generating energy in the core of the Sun shut off and the star starts to collapse. The collapse, however, causes some of the leftover hydrogen gas outside the core to heat up enough that it can begin fusion. Not that there are any slow kinds of fusion, but this fusion happens quickly. The star's core is still collapsing; the star's luminosity, which had been stable throughout its time on the Main Sequence, starts to increase. This increased light makes the outer envelope of gas surrounding the star expand, and the star becomes a subgiant before settling into its new role as a red giant star.

TIP

Red giants are located on the middle-right of the H-R diagram, just above the Main Sequence.

Why are red giants red? As a star expands out into a red giant, its increasingly large surface area causes its total luminosity to also grow. The energy is also heating a larger volume, so it is cooler and redder.

REMEMBER

Red hot? Not! On the color spectrum, red corresponds to longer wavelengths of light which are cooler because they are lower-energy.

The next phase of stellar evolution depends closely on the mass of the star. If a star is much more massive than our Sun, it will continue expanding into what's called a supergiant. If the star is big enough, it can even start a new phase of fusion in its core where it is fueled by helium fusion (or even heavier elements, depending on the size of the star). As each new type of fuel runs out, the star's core can collapse a bit, heat a bit, and begin fusion again with heavier elements. This process, called nucleosynthesis, is how all the elements up to the mass of iron in our universe are formed.

Although it may seem like stars last forever, they don't. Despite many rounds of core fusion, eventually stars run out of fuel. What happens next will depend on — you guessed it — the star's initial mass. Relatively low mass stars (up to about eight times the mass of our Sun) shrink their cores down into what's called a *white dwarf star* (we go into more detail on dwarf stars in Chapter 8), and what's left of the outer layers will be ejected into clouds of gas and dust that forms a planetary nebula. (Can't wait? Skip ahead to Chapter 6.) White dwarfs are located at the bottom left of the H-R diagram, and red giants are at the middle-right, just above the Main Sequence (see Figure 5-5).

Some planetary nebulae look like rings, like the famous Ring Nebula (which is actually a sphere), and others have more complex shapes like hourglasses. Their names come from how they look vaguely round through a small telescope — like Uranus. Looks can be deceiving; planetary nebulae have nothing to do with planets — and they can be one light-year across!

If a star's mass is more than eight times the mass of our Sun, it goes through the same initial steps but much faster, with multiple rounds of core fusion leading all the way up to iron. The final structure of such a star is an unstable core region, with iron being created in the middle, and then rings where silicon, oxygen, neon, carbon, helium, and finally hydrogen are burning as fuel from inside to outside (see Figure 5-5). It will meet an explosive end as a supernova — one of the most sensational endings for anything, ever. More on that in a minute.

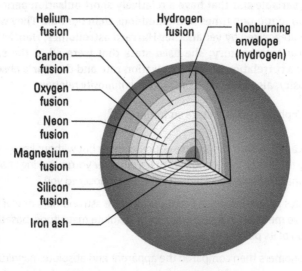

FIGURE 5-5: The interior structure of a pre-supernova high-mass star.

Labels: Helium fusion, Hydrogen fusion, Nonburning envelope (hydrogen), Carbon fusion, Oxygen fusion, Neon fusion, Magnesium fusion, Silicon fusion, Iron ash

The only constant is change with variable stars

Einstein once said that the ability to change is a key indicator of intelligence. By that logic, variable stars are the smartest things out there. Although every star goes through a cycle of burning through its lifetime fuel supply, some stars visibly change in brightness over a period of anywhere from seconds to years. Stars that undergo this kind of observable change are called *variable stars*. They have been observed as early as the 16th century; one of the earliest observed changes in a star's brightness was that of Omicron Ceti, later named Mira, and this finding was incredibly significant in our early understanding of stars as constantly evolving entities.

TECHNICAL STUFF

Case in point: Even our Sun is a variable star. Every 11 years our Sun's output of radiative energy changes by about 0.1 percent — doesn't seem like a lot, but it is measurable.

Pulsating variable stars are a special type of variable star where the contraction and expansion of the star's surface layers cause regular, or pulsating, changes in luminosity. Some variable stars are actually two stars! One star eclipses the other (that is, they are part of a binary star system — we go into more detail in Chapter 6). As an example, imagine two children stepping in front of each other to be first in line for the ice-cream truck. One steps in front to block the other, then the second hops in front of the first in a cycle. Now you see me, now you don't! These are called *eclipsing variable stars*.

Head and shoulders above the others, *Cepheid variables* are an important kind of pulsating variable star that have a relatively short pulsation period (1 to 100 days) but are 200 to 100,000 times more luminous than our Sun. They were first discovered and categorized by yet another Harvard astronomer, Henrietta Swan Leavitt, in the early 20th century; she used stars that were all at the same distance to determine a correlation between pulsation rate and the star's absolute magnitude or luminosity, also known as the *period–luminosity relation*.

Why are Cepheid variable stars so important?

>> You can observe a star's apparent magnitude but without knowing how far away the star is, there's no way to tell whether you're looking at a luminous star that's far away, or a dimmer star that's closer to you.

>> Leavitt's period-luminosity relationship gave astronomers one of the first reliable methods to figure out a star's absolute magnitude, based on the period of its pulsations.

>> Astronomers then compared the apparent and absolute magnitudes, and could get an idea of the star's distance away from the Earth!

>> Such measurements were one of our first ways of figuring out just how big the galaxy, and universe, really are.

>> Cepheid variables are a very important "standard candle," or an object with a known luminosity.

All Good Things Must Come to an End

Whether through a fiery explosion or a slower fizzle back into primordial gas, the life of every star eventually ceases. Of course, the life spans you're looking at aren't short by human standards; the largest stars live 2 to 3 million years, and ones with the smallest masses may live for billions or thousands of billions of years. Do all stars fade quietly into the cosmic sunset? Not even close!

From flare ups to full-blown explosions: Novae and supernovae

Although some stars do go quietly into the night, some end in a flash of brilliance called a *supernova*. *Nova* means *new*, and a nova is a suddenly appearing, bright light in the sky (not all are super). And not all novae are the same; there are several different types of novae. The first occurs in a binary star system (double down on the details in Chapter 6), which can often contain a less-massive and more-massive star orbiting each other. If the more massive star evolves into a white dwarf while the smaller star is still burning on the Main Sequence, hydrogen gas from the regular star ends up on the outside of the white dwarf — the perfect condition for an explosion! This traveling hydrogen gas can ignite on the hot white dwarf's surface.

TECHNICAL STUFF

Because this hydrogen is burning on the outside of the star rather than in the core, the process is unstable and quickly expands to form a shell of hot gas surrounding the white dwarf. To observers on Earth, this nova (or explosion, to the pyromaniacs out there) suddenly appears as a new bright "star" and then fades away, only for more hydrogen to build up on the surface of the white dwarf until it produces another nova burst.

With each nova, the mass of the white dwarf grows a little. If enough mass builds up on the white dwarf through this or another mechanism, the star can instead undergo a sudden collapse. In this runaway process, the star can no longer support itself against all the mass and collapses. In the hot dense remnant, fusion can take place all the way up to the element nickel, which is a little higher than iron on the periodic table. This explosive fusion creates a Type Ia supernova, and the original white dwarf star is completely destroyed in the process.

A separate kind of supernova explosion can take place when the core of a much more massive star runs out of fuel; this explosion produces a neutron star or a black hole (we dive into those in Chapter 8) as the end result, and is called a Type II supernova. Type Ia supernovae are more luminous than Type II, but Type II supernovae can produce much heavier elements — all the way up to gold and even uranium on the periodic table. The superheated gas in the explosion carries these heavy elements out into space and eventually expands and cools, becoming a supernova remnant. In systems with few heavy elements, stars 130 to 250 Solar masses can also explode and leave nothing behind in Type Ic explosions. There is also a neutron star creating Type III supernovae that collapses through a weird process called electron capture.

A quick review of the forces at play with supernovae:

>> Every star is a balancing act of forces.

>> Internal gravity wants to pull the star's core inward.

>> Nuclear fusion at the center wants to push that core outward.

>> After the fuel runs out and temperature drops, pressure at the core decreases.

>> Who wins the force war? Gravity! The results are implosion and/or explosion.

Supernova explosions take place in a galaxy about every 50 years — not every day, but you may be lucky enough to live through one or two in your lifetime. How bright is bright? A supernova explosion doesn't last long but at its peak, it can appear brighter than the galaxy it lives in.

Nucleosynthesis and the creation of new elements

So how are all these star explosions relevant to life on Earth? Short answer: nucleosynthesis tells us that they are absolutely essential. *Nucleosynthesis* is the process that allows for the creation of larger atomic nuclei, or the central cores of the atom, from smaller nuclei. The very first known incidence of nucleosynthesis in the universe came from none other than the Big Bang (flip to Chapter 13 for a preview). After the glorious kickoff of the Big Bang, after the first stars were formed, hydrogen atoms at the cores of stars fused together at incredibly high temperatures and pressures to create helium, just like the Main Sequence stars we talked about earlier in this chapter. The larger the star, the higher the atomic number of elements that can be created through nucleosynthesis. At the same time, smaller elements continued to be combined into larger elements.

A star can progress through burning hydrogen to helium, helium to carbon, carbon to oxygen, and so on through neon, magnesium, silicon, and finally iron. The chain of nucleosynthesis in a star's core stops at iron because although all the smaller elements give off energy when combined, you have to add energy to fuse iron atoms . . . and so it just doesn't happen. Why, you ask?

TECHNICAL STUFF

The mass, say, of one helium atom is less than the mass of the hydrogen atoms that were combined to make it, but this isn't the case for fusing iron Rather than creating energy, iron fusion requires energy. Because the star doesn't have random excess energy to fuse iron, this stops the progression of nucleosynthesis cold. When a high-mass star reaches iron nucleosynthesis and sufficient iron builds up in the star's core, the lack of fusion causes temperature reduction and its resulting instability leads to a Type II supernova as discussed above.

The shockwave of a supernova explosion expands from the core of a star out through its outer layers, producing immensely high temperatures (excess energy, away!) that can result in the nucleosynthesis of elements heavier than iron — all the way up to large atoms like uranium! The spectral signatures of these heavy elements can even be detected in telescopic observations of supernovae.

The life cycle of a star: Heating, cooling, and everything in between

The H-R diagram is useful to compare the mass and luminosity of different types of stars, but it can also be a way to chart the lifetime of a particular type of star. To be clear, because stars live for millions or billions of years, you can't actually use an H-R diagram to follow the evolution of any one star. Instead, astronomers use a star's position on the H-R diagram combined with theory to figure out where it is in its lifetime.

REMEMBER

After stars start nuclear fusion, they have an initial position on the H-R diagram based on their mass and luminosity. From there, our beloved stars have several possible outcomes as they move to the next phase of existence. Stars that are a lot more massive than our Sun tend to evolve into red supergiant stars, and often go out in the incendiary blaze of a supernova that leaves behind a black hole or a neutron star. Stars that are more Sun-sized typically phase into red giants before their final white dwarf stages.

Theorists often plot the future of a star based on its mass on an H-R diagram. For example, in Figure 5-6 you can see the predicted future path for two stars, one with the mass of our Sun (also known as a *solar mass star*) and one that's much larger at 15 times the mass of our Sun.

>> The solar mass star starts in the middle of the Main Sequence and when it exhausts its hydrogen fuel, it follows the red giant branch up and to the right.

This pathway means that its luminosity increases as its surface area grows larger, and its surface temperature decreases.

» The 15-solar-mass star, however, takes a more horizontal branch to the right at the very top of the H-R diagram as it burns through helium, carbon, and oxygen as fuel.

This evolution means that the star's luminosity stays about the same, but its surface temperature decreases as it moves to the right. Look out, folks, we have a supergiant star!

» Both stars in this case eventually reach an end state either as a white dwarf (in the case of the solar-mass star) or as a supernova and perhaps a neutron star or black hole.

More on these objects in Chapter 8.

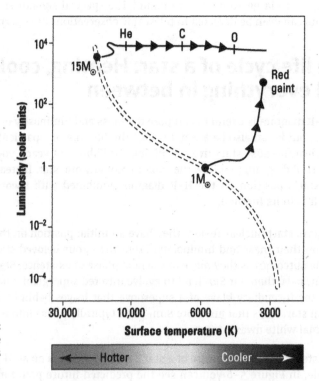

FIGURE 5-6:
The possible life cycle phases of a star.

Star reincarnation and the next phase of stellar existence

Elements up to iron are created in the cores of stars, and heavier elements are created in supernova explosions and other weird events. What happens to those elements? Consider them as being reincarnated as future stars and planets. The very first stars began with only hydrogen as fuel. Stellar explosions resulting from the death of these stars led to the creation of second-generation stars. These stars contain more metals than first-generation stars because they are composed of first-generation leftovers, including metals made through nucleosynthesis. Third-generation stars created from these explosions have even more metal than earlier-generation stars. You can actually observe whether a star is metal-rich or metal-poor, using spectroscopy, and get an idea of when that star might have formed. See Chapter 14 for more on these different generations of stars, called *Populations*, and how they formed following the Big Bang.

Let's use our Sun as a case study. The Sun is a middle-sized dwarf star and based on its mass, the highest levels of nucleosynthesis that it will ever be able to reach are the creation of carbon and oxygen in its core. It will, for example, never be able to make its own iron. But when you look at the Sun's spectrum using dedicated solar telescopes (either earthly or space-based), you can see all sorts of more exotic elements — nitrogen, sodium, silicon, magnesium, and iron have been detected, as well as rare elements like vanadium and europium. Because these elements couldn't have been created by the Sun on its own, we can surmise that they formed in ancient supernovae.

The materials from supernovae early in the galaxy's history would have been spread out through space and eventually ended up as the molecular cloud of gas and dust from which our Sun was born, a mere 4.6 billion years ago. Because the Earth and the rest of the solar system formed from the material in the protosolar nebula, our iron and other heavy elements also came from those same supernovae explosions (and other weird events) billions of years ago.(For example, neutron star mergers make gold!)

TIP

Although you can mine for iron on Earth, the Earth itself cannot create heavy metals. The metals we mine and use are part of the primordial material the Earth was created from, and this material was actually created from the stars.

All of the elements heavier than helium in the human body were also created in the explosion of a dying star. If you're ever pondering the existential nature of where you came from, now you know. Someday, when the Sun reaches the end of its lifetime, its elements and those from the Earth and the rest of the solar system may be dispersed through space, only to one day become part of a new star. Perhaps a form of reincarnation, though not in any form you've likely ever imagined!

Chapter **6**

Friends for Life: Star Systems and Dust Clouds

ndividual stars have a creation story all their own. Once upon a time, a cloud of dust and hydrogen gathered. Although these elements put up a good fight, they eventually succumbed to gravity. Compacting and collapsing into a protostar, our star-in-the-making finally evolves into its not-so-final form as a bona fide star. As these happy-go-lucky stars live their lives (check out Chapter 5 for the full story), they gradually burn through their hydrogen supply and undergo nuclear fusion; from this point, all stars reach a decision-making point based mostly on their mass. Some fade out of existence, whereas others go into the night less quietly by collapsing and exploding as a supernova.

REMEMBER

Not every light we see in the sky is a star. Other objects visible with the naked eye include planets such as Venus and Jupiter, our Moon, the occasional comet or meteor, and even satellites such as the International Space Station.

Although stellar life and death are great milestones, what happens in between those milestones is also very interesting. You can often see individual stars when looking up into the night sky, but did you know that most of those are actually part of multiple-star systems? Up to 85 percent of larger stars belong to multi-star systems. These systems could be binary (two-star systems) or triple-star systems containing three stars whose orbits are gravitationally bound together, and the list goes on as more stars join the gravitational party. And stars aren't the only thing in the sky — clusters, nebulae, and other aggregate astronomical objects require a deeper understanding of astrophysics, one that will open your mind to the myriad possibilities in our own galaxy.

The More the Merrier: Binary and Multiple Star Systems

It takes two to tango, and stellar dancing becomes a complex affair involving a third party: gravity. Individual stars rotate on their axes in addition to orbiting around the center of the galaxy. Life sure was easy when it was just one! With binary star systems, two stars are gravitationally bound to each other and they orbit around their common center of mass, the *barycenter*. Just to make things more interesting, binary stars aren't twins or clones of each other. In fact, it's just the opposite — binary stars may be wildly different from each other in terms of their mass and evolution. Adding more than two stars in a system makes the system more intriguing from an academic point of view, but also much more complicated! And to further increase the complexity, it's not always obvious whether two stars just appear next to each other, or if they are an actual multi-star system. We talk about how to tell the difference in a bit.

REMEMBER

Need a quick way to remember orbital period, or the time required for one astronomical object to fully orbit another? Every calendar year the Earth orbits the Sun, making the Earth's *orbital period* one year. Orbital period isn't the same as *rotational period*, or the amount of time it takes for an object to rotate once on its axis. Earth's rotational period is one day compared to the Sun.

The ties that bind: What it means to be gravitationally bound

You can't have a binary star system without two stars orbiting each other, and two stars will not mutually orbit without gravity. As discussed in Chapter 2, the gravitational force is responsible for physical attraction between any two objects. Pick up a pencil and drop it; gravity is the force that pulls that pencil down toward the center of the Earth (and, inexplicably, always somewhere you cannot reach). Objects are said to be *gravitationally bound* when they're physically near enough to each other that gravity keeps them together, and the gravitational force is proportional to the masses of the objects and the distance between them.

The gravitational force between two objects is

$$F = \frac{Gm_1m_2}{r^2}$$

In this equation, F is the gravitational force, G is the gravitational constant, m_1 and m_2 are the two masses, and r is the distance between them. You can see from this equation that the force is directly dependent on the two masses, and inversely dependent on the distance — so the closer the objects are to each other, the greater the force.

Gravitational forces can also produce some special locations in a binary star system (or any system where one object orbits another one). Picture a binary star system consisting of Star A and Star B, and draw a straight line between them. At a point on the line close to Star A, its gravitational force pulls more strongly than the force from Star B, and the same applies if the point is close to Star B. But because the gravitational force decreases as the square of distance decreases, at some point in between Star A and Star B these forces will match. The location of this point will depend on the masses of the two stars — if the masses of Star A and Star B are equal, then this point will be exactly halfway in between the two stars. But if Star A is more massive, the location of this point will be closer to Star B because the pull from Star A will be bigger. A different special point is the center of gravity, which is called the *barycenter* (see Figure 6-1), and it's also the point around which the two stars orbit. The center of gravity is the point you could balance the solar system on, and it will be closer to the larger star or maybe inside it.

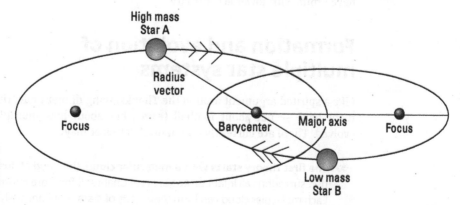

High mass
Star A

Radius
vector

Focus

Barycenter

Major axis

Focus

Low mass
Star B

FIGURE 6-1:
Diagram showing
orbits in a binary
star system.

Mapping out the shape of where the gravitational force on an imaginary particle is constant (not just on a line, but in dimensional space surrounding the star) creates what's called a *Roche lobe*. Binary star systems contain two Roche lobes, one each surrounding Star A and Star B. These lobes are shaped like teardrops; the point where they touch is where the forces balance, and the balance point is called the first Lagrange point (abbreviated as L1).

Orbits and the science behind them

Ring around the rosy? Close, but not quite. Everything in space is in constant motion, and astronomical objects such as stars, planets, and moons enjoy the benefits of gravitational binding. An *orbit* is a repeating path or trajectory (curved, when you're talking about the cosmos) that one object follows around another. Most orbits are not perfectly circular! Kepler's laws of planetary motion (Chapter 2 contains the details) showed that planetary orbits are shaped like ellipses. Although Kepler was thinking about the shape of an orbit of a planet around our Sun, binary star systems in which two objects orbit around each other also include ellipses. In Figure 6-2, the orbits of Star A and Star B are shaped like overlapping ellipses with the barycenter at the shared focus.

REMEMBER

A circle is a special case of an ellipse where the two foci are co-located at the center. The circle scenario is called *zero eccentricity*. Eccentricity measures the non-circularity of a star's orbit as a value between 0 and 1; higher values mean a star's orbit is more elliptical than circular. You can calculate an object's eccentricity (refer to Figure 6-1) using the formula $e = c/a$, where e = eccentricity, c = the distance between the ellipse's focal points, and a = the length of the ellipse's major axis. Some binary star systems have highly elliptical orbits, whereas others may have orbits with lower eccentricity.

Formation and evolution of multiple star systems

Like a spirited argument around the Thanksgiving dinner table, there's disagreement here — not about football teams, but about how multiple star systems evolved. There are currently two main theories at play:

>> **The first theory starts with a molecular cloud,** the cloud of dust and gas that surrounds all infant protostars (see Chapter 5 for more information). Each molecular cloud can form thousands of stars simultaneously.

As gas and dust clump together and become gravitationally bound, some of the cores form closely enough that their gravity affects each other.

Although the cores are far enough apart to avoid merging into a single larger star (at least not right away) the gravitational tugs from each protostar cause them to orbit around each other, eventually becoming a binary star system after both ignite.

>> **A second theory is a tad more cannibalistic.** It's possible that when two or more individual protostars are close enough, one may encapsulate the others within its orbit and lead to a multiple star system.

The difference between this theory and the first is that in this version, the protostars ignite into actual stars first, and then capture each other to form a multiple star system. It's possible that both are true!

Interestingly, binary star systems can also affect each other's evolution by transferring material between them. Remember those Roche lobes? It turns out that if you have material orbiting one star (or even part of one star) and it gets to the Lagrange point L1, that material can transfer directly to the other star. In the case of binary star systems, if Star A and Star B are smaller in size than their Roche lobes, only material in orbit can be transferred. This type of system, called a *detached binary,* is most common. Now suppose, for example, that Star A expands to fill out its entire Roche lobe. In this case, the shape of Star A stretches to the point where material can transfer over to Star B across L1. This variation on the binary star theme is called a *semi-detached binary* — the two stars are still separate, but mass moves from one to the other.

If both Star A and Star B fill out their Roche lobes completely, what's left? A *contact binary system,* that's what. In this type of system, two stars are essentially "touching" and they are surrounded by an outer cloud of gas that envelopes both stars but isn't bound just to one of them. In this case, both stars will have a distorted, teardrop shape or they will share a single flattened sphere envelope.

The cosmos is chock-full of other weird binary star systems that include mass transfer from one star to another. The white dwarf mass transfer from Chapter 5 that resulted in a Type Ia supernova is in this category. Another notable example is x-ray binaries, where material is transferred from a normal main sequence star onto a more evolved, super-dense companion star like a neutron star or even a black hole. As the material falls onto the super-massive companion under the attraction of gravity, it is accelerated and super-heated until it gives off high-energy x-rays. X-ray binary star systems are further subdivided into categories based on mass and the ratio of energy emitted as x-ray versus visible light.

TECHNICAL STUFF

One particularly interesting subcategory is called an *x-ray burster.* Rather than a constant source of x-rays, these bursters release — well, for lack of a better term — bursts of extremely high energy x-ray radiation, which then diminish over the course of seconds or minutes. The bursts usually recur on a timescale of hours or days. The physics of x-ray bursters is similar to what happens with classical novae when material piles up on a white dwarf, but x-ray bursters involve a neutron star rather than a white dwarf.

There are a number of other kinds of binary star systems, mostly named for the method used to detect them. Here's a short list; see Figure 6-2 for a summary.

Types of binary star systems:

>> **Double star:** Two stars that appear next to each other to the unaided eye, or through a small telescope; may not be an actual binary star system.

>> **Visual binary:** Binary system located close enough to us that we can see the individual stars. We can track their positions over time to figure out the orbital period and mass of the stars.

>> **Spectroscopic binary:** Binary system detected through Doppler shifts in spectral lines; method used to detect the majority of binaries.

>> **Eclipsing binary:** Binary star system in which one star passes in front of the other as seen from Earth, causing the total brightness to dim periodically.

>> **Astrometric binary:** Binary star system detected through the wobbling motion of one star, used to detect an unseen companion. Requires long-period observations.

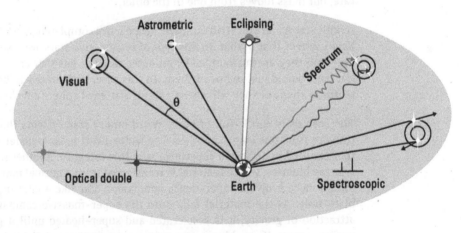

FIGURE 6-2:
Types of binary star systems.

Step into the light: Eclipses in a double star system

Eclipses are the subject of multiple songs, witty sayings, and school science fair projects.

REMEMBER

In general terms, one celestial body obscuring another is defined as an eclipse; eclipses occur when one body (typically the Sun or Moon, from our Earthling perspective) passes into the shadow of another (see Chapter 3).

An eclipsing binary star system is one where one star passes in front of the other as seen from Earth. When the orbital plane of the binary star system is positioned edge-on, you can view the stars' eclipse at two points in their orbital period — once for each star in the system. Astronomers use photometry to detect an eclipsing binary star system; photometry is a way of measuring the brightness of a star to very high precision. If a star's brightness changes over time in a periodic way, as compared to the other stars near it, either it's a variable star (see Chapter 5) or an eclipsing binary. This technique works for stars that are too close to each other to resolve into individual stars with our telescopes on Earth, so we can only measure the combined brightness of both stars.

Astronomers use what's called a *light curve* (see Figure 6-3) to measure eclipsing binary star systems.

TIP

A light curve is nothing more than a graph comparing brightness to time. Observations taken over many weeks or months are combined to look for periodic variations.

TECHNICAL STUFF

Usually, a light curve for an eclipsing binary star system will have troughs which alternate in depth — these dips are called the primary eclipse and the secondary eclipse. In most binary star systems, one star appears brighter than the other. If Star A appears brighter than Star B, Star B's passing in front of Star A causes the total brightness of the combined system to dip. This dip shows up as the primary eclipse on the light curve. When Star A passes in front of Star B, less light is lost proportionally, and the light curve dips less; this lesser dip would be the secondary eclipse. Studying light curves allows you to figure out the relative brightness of the two stars in an eclipsing binary system, and also provides information about the system's orbital inclination and the sizes of the stars themselves.

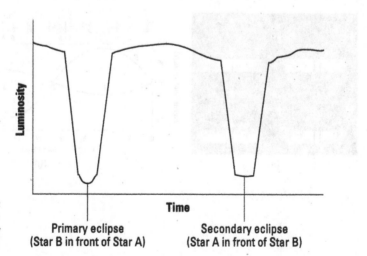

FIGURE 6-3:
Light curve of an eclipsing binary star system.

Binary star orbits and radial velocity

Eclipsing binaries are amazing to witness, but the dips in brightness are only visible if the star system happens to be aligned edge-on as seen from Earth. Given that not all binary stars are willing to cooperate with our paltry Earth-based viewing, an alternative method for isolating an observational double star from a true binary system is by using radial velocities. This technique is also used to detect planets orbiting other stars (more on this in Chapter 7). Remember that each star gives off its own unique spectral signature. Star light emissions can be spread out into their many wavelengths using a special instrument on a telescope. Dark or bright lines correspond to diagnostic wavelengths for different chemical compounds in a star's atmosphere; Doppler shifts (see Chapter 4) can also be used to learn more about the motions of the two stars.

Imagine the scenario where two stars in a binary system orbit their common center of mass, one star heading towards us (the lines in its spectrum will be blue shifted) and the other heading away (its lines will be red shifted).

TECHNICAL STUFF

This motion toward or away from us is called the *radial velocity*. When you observe the stars' combined spectrum, the spectral lines will look doubled.

Spectral changes can be seen by observing a star over a period of weeks, or months, or even years in some cases, to look for the diagnostic signature of a binary star system as the spectral lines move apart and then back together; see Figure 6-4 (left). You can then make a plot of radial velocity (see Figure 6-4, right) and look at the time period over which the radial velocities of the two stars change. Congratulations, not only have you found a binary star system, you've also figured out its orbital period!

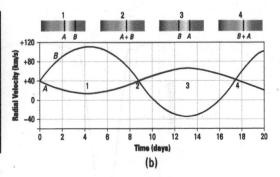

FIGURE 6-4: Binary star system spectrum and radial velocity.

But wait, there's more! After you've figured out the period of a binary star system using the preceding radial velocity technique, you can even figure out the masses of the two stars. Sir Isaac Newton (famous for using a falling apple to develop a theory of gravity) took Kepler's third law of motion and rewrote it for two objects in orbit around their mutual center of mass. Case in point: a binary star system. In this case, we have

$$D^3 \infty (M_1 + M_2)P^2$$

where D is the size of the orbit, P is the orbital period, and M_1 and M_2 are the masses of the two stars. This equation shows that the orbital period is defined by the total mass and the size of the orbits.

Knowing the speed of the stars through redshift combined with their period from the velocity curve enables you to figure out the size of the orbit based on how fast the stars are moving and how long they take to go around the orbit once (the *period*). That's enough information to plug into the equation above, and presto — you've got the sum of the masses of the two stars. Relative orbital speeds will then yield the individual stars' masses.

This method makes some assumptions about the orientation of the star system with respect to us on Earth (such as, how tilted its orbit is compared to ours), but this method also doesn't require knowing how far away the stars are.

Three's not a crowd: Multi-star systems

If two's a crowd, three's a party! Binary star systems involve two stars each, can be composed of entirely different kinds of stars, and can even eclipse each other throughout their orbits. But what happens if another star wants to join in the fun? Multiple-star systems have an inherently complex physical nature because their orbits don't cleanly align around the common center of mass as they would with binary stars.

Take a look at how multiple star systems exist. The main scenarios are

- » Two stars may share an orbit with a third (or fourth, or fifth) star orbiting that pair.
- » Two pairs may orbit each other as part of the same system.

Binary star systems are relatively stable in that they orbit a common center of mass. You only got have orbits to worry about, after all! With systems of three or more stars, chaotic behavior may ensue because there's no single, clear path for predicting the stars' orbits because one star can have a compounding effect with the other stars in the system — fun to think about, but impossible to predict in all but special situations.

TECHNICAL STUFF

The inherent confusion created by these forces and orbits is described scientifically as the "Three Body Problem," one of the oldest scientific dilemmas that has yet to be generally solved. It's an example of a chaotic system that can't be fully predicted, even by our best computers — all astrophysicists can do is to model the probability of any potential outcome.

Some cases of multi-star systems are more understandable. Consider a binary star system that's relatively closely clustered, combined with a third star that orbits the two at a much farther distance. Because the central binary star system is much closer together than the third star, the third star orbits the mutual center of mass of the binary. Systems like this can get more and more complicated — there's even a six-star system called TYC 7037-89-1! This system consists of three pairs of binary stars, two of which orbit each other. The third binary pair then orbits around the other two pairs. See Figure 6-5 for more info on this complicated sextuple dance. Table 6-1 contains examples of some well-known multiple star systems.

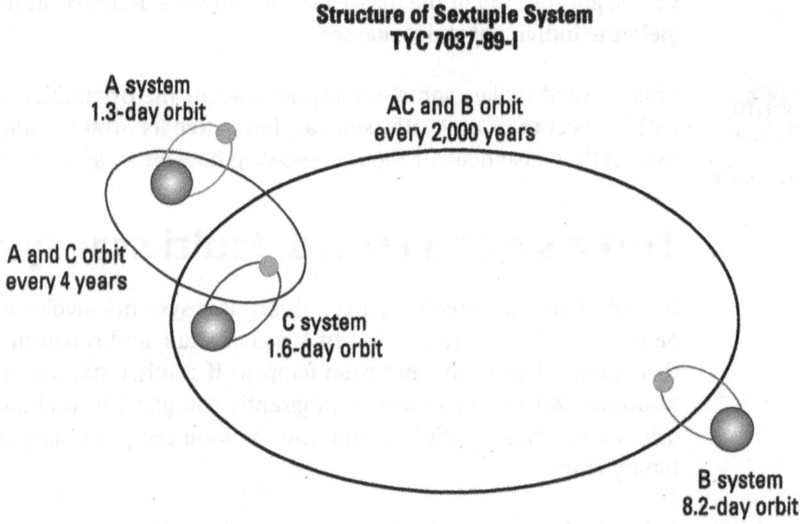

Structure of Sextuple System
TYC 7037-89-I

A system
1.3-day orbit

AC and B orbit
every 2,000 years

A and C orbit
every 4 years

C system
1.6-day orbit

B system
8.2-day orbit

FIGURE 6-5:
The orbital paths of the six stars in the TYC 7037-89-1 system.

TABLE 6-1 ## Quick Differences between Types of Star Clusters

	Open Star Clusters	Globular Star Clusters
Typical number of stars	Up to a few thousand	Hundreds of thousands
Typical star age	Hundreds of million years old, up to billions	10–12 billion years old
Density	Loosely packed	Highly dense
Shape	Irregular	Symmetrical, round
Size	Diameter < 30 light-years	Diameter < 300 light-years

Huddle Up There, Star Clusters

Individual stars have their own creation stories, properties, and lifestyles. Binary stars are best friends for life (or close to it), and multiple-star systems add several more stars to the gravitational mix. What do we call groupings of hundreds, or even thousands or millions, of stars that are bound to each other through gravity? You guessed it — *star clusters*.

Getting sticky with cluster formation

Molecular clouds of dust and gas compress and condense (see Chapter 5); as the gravitational forces acting at the core vacuum up nearby material, a star is formed. If enough stars are close enough to each other to be bound together gravitationally, you have a star cluster.

TIP

What makes a grouping of stars a cluster and not a galaxy? The answer to this question lies in the presence or absence of dark matter, covered in Chapter 15. If the cluster of astronomical objects contains dark matter, there's a good chance it's a galaxy; without dark matter, it's more likely a star cluster.

Shut the door! Learn about open clusters

Star clusters' degree of "stickiness" varies between clusters. Perhaps the least gluey members of the star cluster family are open clusters. These loosely clustered groupings of stars can have tens to hundreds of stars. They can be located within irregular and spiral types of galaxies (more on this in Chapter 9), and because they are only loosely bound, stars can race ahead or lag behind as the cluster orbits our galaxy. Eventually, our crowd of racers becomes a thousand separate stars. Open clusters contain relatively young stars, and age really is relative in the cosmos — most stars in an open cluster are tens to hundreds of millions years old.

TIP

Astronomers have found over a thousand open clusters so far, mostly in our galaxy's disk and some in nearby galaxies. Some open clusters are young enough to still be embedded within the nebula they formed from, and many consist of young bright stars that burn with a blue color.

One famous open star cluster is the Pleiades (Messier number M45, or Subaru in Japanese), a visible cluster of at least six bright stars in the constellation Taurus (see Figure 6-6). If you look at the Pleiades with binoculars or a small telescope, you see that it's made up of dozens of stars. It's located about 440 light-years away from us, and scientists think the stars are about 100 million years old.

TECHNICAL STUFF

Because the stars in an open cluster formed at the same time, they are all the same age.

Get even stickier with globular clusters

At the other end of the star cluster spectrum, we invite you to meet up with globular clusters. Named as such because they are globe-shaped, these star clusters are generally much larger than open clusters; they can contain up to millions of stars and are grouped more densely than open clusters. They are also much older than open clusters; stars in globular clusters tend to be up to 10 to 12 billion years old. Yes, billions! Globular clusters can be found in a sphere around the Milky Way Galaxy (there are at least 150 around the Milky Way) and in other galaxies. It's not fully understood how globular clusters form, but they form all at once like open clusters.

Globular clusters are particularly interesting because they are so tightly bound together that they tend to be stable over long periods of time — more loosely bound open clusters can drift away more easily. Globular clusters have a distinctive spherical shape and are closely packed. The Hubble Space Telescope came in handy for discerning individual stars in star clusters and has been used to finally shed light on how they form and evolve. Figure 6-7 shows a tightly-packed ancient globular cluster, Messier 15 (M15 for short), which is in the constellation Pegasus and is about 12 billion years old.

Globular clusters are also the home of some weird objects. There are white dwarf stars, but also neutron stars and the occasional smaller black holes — scientists think that when these merge together, they could be a source of

FIGURE 6-7:
Image of globular
cluster M15, as
seen by the
Hubble Space
Telescope.

Courtesy of NASA, ESA

gravitational waves (more on these in Chapter 8). They are also home to stars called *blue stragglers*, old stars that mysteriously get reinvigorated and start looking and acting like younger stars. Blue stragglers seem to either drag fuel off a nearby star or get it through a collision, and suddenly start burning hotter and brighter.

Pedal to the Metal with Interstellar Gas and Dust

When the kids are hungry and the fridge is bare, resourceful parents scavenge the pantry and come up with something edible — and oftentimes delicious. When you've forgotten a coworker's birthday, you might borrow a rose from a neighbor's bush and leave your colleague a birthday bouquet. We humans have an innate ability to create something from nothing. Perhaps the source of this spirit of ingenuous creation is our ancestors from afar, the stars.

The Interstellar medium

You're now familiar with stars being created from galactic clouds of dust and gas — it would be hard to miss these, because we've said "dust" and "gas" more times than you probably care to count (and please don't count . . . we repeat these many more times over the course of this book). After that gas and dust condenses and forms a star, what are we left with? The space between stars is known as the "interstellar medium." Contrary to popular fiction, space is not a perfect vacuum. Low-density dust and gas remain in between stars, and comprise up to 10 to 15 percent of the "empty" space of a galaxy such as the Milky Way. The main composite elements of our interstellar molecular gas are hydrogen and helium (about 98 percent of the total makeup), along with smaller amounts of heavier elements such as nitrogen and carbon. These atoms are mostly in the form of neutral atoms and molecules, but the hot areas of the interstellar medium also has some charged particles such as ions and electrons. It's super-low density, though, averaging only about one atom per cubic centimeter! For comparison, the Earth's atmosphere has around 3×10^{19} molecules in a single cubic center at sea level. The interstellar medium may not be a complete vacuum, but it's pretty close.

Dreaming with your head in the (molecular) clouds

The interstellar medium take many forms, and even individual structure can vary wildly in temperature and density. In the zoo of structure, one of the most foundational elements is the molecular clouds (see Chapter 5).

REMEMBER

Molecular clouds earned their name because most of their hydrogen is in molecular form as H_2. Some of the more active regions for star formation are in molecular clouds. They are also the densest and coldest types of interstellar cloud — close to absolute zero. They vary widely in size, between one and hundreds of light-years in diameter, and can provide sufficient birthing materials to hundreds and thousands of stars. Although they are mostly made of hydrogen, they do have other molecules in them; astronomers can use masers to figure out their composition.

IR spectroscopy and laboratory astrophysics

A further dive into the nitty-gritty of the interstellar medium reveals the role of laboratory astrophysics, or what we can learn about interstellar space via advanced tools.

REMEMBER

IR radiation, or infrared radiation, is the portion of the electromagnetic spectrum where wavelengths are longer than what you can see with your eyes; although infrared radiation is invisible, you can feel its energy in the form of heat.

IR spectroscopy is a method of analyzing infrared radiation and its interaction with matter. These interactions are quantified by directly measuring how a molecule reflects, emits, and absorbs light.

Why do we need IR spectroscopy to study the space between stars? If you have a molecular cloud or other distant object and you want to figure out its composition, you can use IR observations. Specifically, this technique uses advanced chemistry to find out exactly what's occurring in those molecules. Observations of the spectra of objects in space are compared with catalogues of laboratory measurements of different materials. It's a bit of a detective story, but after you find a match for the observed emission or absorption features, you know one of the chemical components.

One problem with infrared observations is that telescopes on the ground must look through Earth's atmosphere, and there's significant absorption in the IR part of the spectrum due to the water vapor in our skies. One way to solve this problem is to have a look from space. The Herschel Space Observatory, for example, was a space telescope that used the far-infrared spectrum to study the chemistry occurring in interstellar space. Herschel was launched in 2009 and remained active until 2013, providing critical information that scientists used in replicating the low temperature conditions that are found in interstellar space. The James Webb Space Telescope also operates in the IR portion of the spectrum.

Adding Structure to That Gas and Dust: Nebulae

You may not be able to see the forest for the trees, but can you pick out a nebula from a sea of particles? Absolutely! Nebulae are another feature of the Interstellar Medium. These clouds of particles are mainly gas and dust (here they are again!), that exist in between stars. Not only are nebulae composed of gas and dust, but some are also created from clouds of gas and dust, and these serve as the birthplace for stars that are created from — you guessed it — gas and dust. Others come from stars dying. They lack the gravitational binding that holds a star together but provide us with many other key data points about the cosmos. We've already talked about a few specific kinds of nebulae, such as molecular clouds and star formation regions, but there's more to these clouds.

TIP

The name *nebula* comes from the Latin word for *cloud*. Refreshing, isn't it, to finally call something what it looks like?

The interstellar medium, you now know, is the stuff in between stars. Enter the nebula, an accumulation of gas (typically ionized gases such as hydrogen) and dust that's denser than the interstellar medium surrounding it. Although a lot of the interstellar medium is dark, many nebulae glow either from star light or because they are hot enough to give off light.

Although many nebulae date back to early in the universe, some have been created more recently. For example, nebulae can also be created from the gas and dust expelled during a supernova explosion.

Nebulae are usually categorized based on how they look, how they formed, or how they are observed.

The main types, or classes, of nebulae are as follows:

>> **Diffuse nebulae:** Includes both emission nebulae and reflection nebulae

>> **Dark nebulae:** Also called *absorption* nebulae

>> **Planetary nebulae:** Oddly, have nothing to do with planets

>> **Supernova remnants:** A form from the exploding star's energy slamming into surrounding interstellar medium that may have been otherwise dark

Please see the color image section for some gorgeous nebula images.

Diffuse and dark nebulae

A common type of nebulae, diffuse nebulae, are clouds of gas and dust without any sharp boundaries. They can be created from leftover material ejected from past star explosions or may simply exist as part of the interstellar medium. So named because they are mostly gas (hydrogen along with lesser percentages of other gases and metals), diffuse nebulae tend to pile up in the spiral arms of their parent galaxy. These prime star formation regions can be further broken down into reflection and emission nebulae.

An emission nebula is hot enough that its constituent gas, mostly hydrogen, gives off energy. Think of an emission nebula as a glow-in-the-dark ball but without the ball; they're created when hot stars in the neighborhood emit enough photons to ionize their gas. A diffuse emission nebula may also be called an H II region — the H is for hydrogen, and the roman numeral indicates its ionization. H I would be neutral hydrogen, and H II is the first degree of ionization (where the hydrogen atoms have lost their only electron).

A reflection nebula is one whose gas isn't hot enough to actually cause ionization. Instead, these nebulae are visible because the light from nearby stars reflects off the dust it contains, giving the reflection nebula a blue color. Because most diffuse nebulae are made up of both gas and dust, many emission nebulae also have a component that's a reflection nebula. Side note: These are absolutely breathtaking when viewed through a telescope.

Diffuse nebulae are often star-formation regions. Much of the time, light from the newly forming stars inside the nebula ionizes the gas (causing emission) and bounces off the dust (causing reflection). As more and more stars form in a star formation region, eventually most of the gas and dust will have been used up by the stars. The rest gets blown away from stellar winds coming off the young hot stars, and what remains is an open star cluster.

Finally, not to be outdone, some nebulae resemble dust bunnies! Dark nebulae contain mostly concentrated dust, so concentrated that it's extremely dense and absorbs the visible light around it. Something that absorbs light is too dark to see, right? Put your detective cap back on; we can observe dark nebulae when they're positioned such that they obscure the light of other objects.

REMEMBER

Never underestimate the power of darkness. The Horsehead Nebula in Orion (see Figure 6-8) is one of the most famous dark nebulae because, as the name suggests, it's shaped like a horse's head. Even though it's composed nearly entirely of dust, a powerful magnetic field forms the dust into its characteristic shape. The red, illuminated hydrogen gas behind it certainly makes identification easier.

FIGURE 6-8:
The Horsehead
Nebula, as seen
by the Hubble
Space Telescope.

*Courtesy of NASA, ESA, and the Hubble Heritage Team
(AURA/STScI)*

Who made who? Planetary nebulae and supernova remnants

Diffuse nebulae aren't the only game in town (er, space!). Planetary nebulae are created when material ejected from a dying star is illuminated by a left-behind white dwarf. They have nothing to do with planets, though they do visually resemble disks like the giant planets.

In Chapter 5, we talk about the H–R diagram and how a star on the main sequence evolves from a red giant to a white dwarf near the end of its lifetime. When an inflated red giant star finally becomes unstable and collapses into a white dwarf, it ejects a large portion of its mass into a tightly bound expanding gas shell surrounding the leftover hot central stellar core — the white dwarf. The core gives off high-energy radiation and has fast outflows of stellar winds that impact with the expanding gas shell, leading to energy that gives off light. Telescopic observations of the spectra of planetary nebulae show highly ionized gases such as oxygen, as well as neon, nitrogen, and other typical elements such as hydrogen and helium. Planetary nebulae are quite short-lived, astronomically speaking — they usually are only visible for thousands to tens of thousands of years before they spread out too much to be visible.

Not to be outdone, stars significantly larger than our Sun can also create their own kind of novae at the end of their lifetimes. Large stars will explode in a violent supernova. Supernovae remnants are created as the aftermath of a supernova explosion. They can appear in different shapes such as the "crab-like remnant" (yes, that's really their name) or the "shell-type remnant."

After a star explodes in a supernova, it produces a hot mess — literally! Gas and dust are thrown out in a chaotic cloud which rapidly expands and fades. This material can travel at incredible speeds, up to 10 percent of the speed of light. Energy from the explosion creates a strong shockwave, ionizing and heating plasma from the surrounding interstellar medium as well as the material from the supernova itself. The plasma can be heated up to temperatures of millions of degrees. Like planetary nebulae, supernovae expand and fade away over time. After 20,000 years they are hard to make out, and after millions of years they will completely mix into their surroundings.

We can use the standard distance formula, $D = RT$ (distance D equals rate R multiplied by time T) to estimate the age of a remnant nebula. Scientists can compare images of these nebulae taken years apart and use that data to calculate the rate of expansion which, in turn, provides us with an estimate of when its parent supernova exploded. Images of the Crab Nebula taken over a couple of decades will do the trick.

Some supernova remnants have a neutron star or a pulsar (spinning neutron star) left in the middle, whereas others end up with a black hole (see Chapter 8 for more detail). These explosions heat up the interstellar medium and spread the heavier elements formed through nucleosynthesis throughout the galaxy, thereby accelerating cosmic rays that can be detected from Earth.

Chapter **7**

Exoplanets: The Search for Earth 2.0

O ver the past centuries, the night sky has gone from a source of wonder and inspiration to a natural system, one that humans have only just begun to understand through observation and science. One of the enduring cosmic mysteries is whether solar systems with planets orbiting a star like our Sun are rare or common. In the past 25 years, astronomers have finally had the technology to detect extrasolar planets, also called *exoplanets,* and the results have been stranger and more wonderful than anyone could have imagined.

The solar system we know and love is home base for our planets: Mercury, Venus, Earth, Mars, Jupiter, Saturn, Uranus, and Neptune, plus countless dwarf planets (hello, Pluto!), asteroids, and other smaller bodies. Do planets exist outside of our solar system? Absolutely! Any planet outside our particular solar system is called an *exoplanet,* and there are at least 5000 that we know about, with more being discovered every year. Could one of those planets be like Earth and maybe even support life? Let's see!

TIP

Keep up with the latest exoplanet discoveries at the NASA Exoplanet Catalog website (https://exoplanets.nasa.gov/discovery/exoplanet-catalog/). It's frequently updated with the latest info on confirmed and candidate exoplanet systems detected with ground-based and space-based telescopes.

Beyond beyond Earth

Before we talk about exoplanets, we make a brief detour through our own solar system. Our planets are classified generally into two groups:

>> **Four inner rocky planets:** Mercury, Venus, Earth, and Mars

>> **Four outer gas giant planets:** Jupiter, Saturn, Uranus, and Neptune

There's a general decrease in density and temperature the farther one gets from the Sun. Mercury is a hot body with a huge metallic core, whereas Mars has a smaller core and lower density. This asteroid belt is next in line; this celestial beauty is composed of leftover rocky and icy material from the early formation of the solar system, and extends out to the giant planets. Jupiter and Saturn have huge atmospheres of hydrogen and helium, whereas Uranus and Neptune have larger percentages of water, ice, and other materials. Around and beyond Neptune is the realm of the Kuiper belt and the Oort Cloud; the small bodies here are made mostly of ice and are primary source regions for comets. Figure 7-1 shows the major bodies in our solar system, in order out from the Sun (not to scale). Both the asteroid belt and Kuiper belt have dwarf planets.

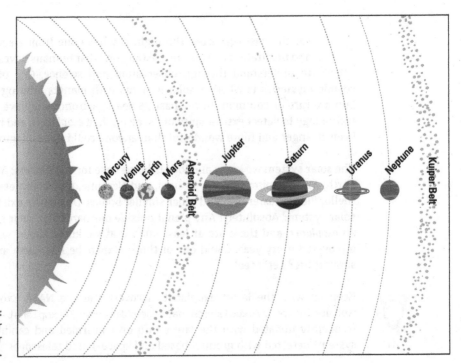

FIGURE 7-1:
The major bodies in our solar system.

Exoplanets explained

Reach beyond our solar system and what do you find? Stars, stars, and more stars . . . and planets! Any planet outside of our solar system is called an exoplanet. And what exactly makes a celestial body a planet? By definition, a planet meets three requirements:

>> Planets are in hydrostatic equilibrium — that is, they're spherical due to their mass and internal gravity.

>> Planets orbit around a star or exist between the stars (they go rogue!).

>> Planets have enough gravitational force to push smaller bodies out of their orbit, except for their own satellites.

The technical definition of our solar system is that it consists of our Sun (our star) and its surrounding gravitationally bound objects (the planets and other smaller bodies).

TIP

Although we now know there are other solar systems out there, "the solar system" generally refers to ours. Exoplanets are any planets outside our solar system, and they were first discovered in 1992 — very recently by universal standards!

Given that we can't fly to an exoplanet and check it out, what do we know about them? Courtesy of the Kepler Space Telescope and other observatories, the vast majority of known exoplanets are within our Milky Way Galaxy — which is not to say that other galaxies don't also have exoplanets. They likely do, but they are much harder to detect. Through calculations of their mass and size, we know that exoplanet terrain varies as widely as planets in our own solar system — they can be rocky, icy, or gaseous. Temperature and size vary as do exoplanet orbits. For more on these topics see the later section "Exoplanets Come in Many Shapes and Sizes."

WHY WAS PLUTO DEMOTED?

If you're of a certain age, you probably learned that there were nine planets in the solar system, not eight. Don't worry, you aren't crazy — in 2006, the International Astronomical Union tightened up its solar system definitions and decided that although Pluto met the first two criteria above (it is spherical and orbits a star), it doesn't meet the third. Pluto and all its moons share their orbit with other objects that orbit the Sun, not Pluto. These objects are called *plutinos* and orbit the Sun within the Kuiper belt. Pluto is now considered a "dwarf planet" instead of a full-blown planet.

Key to understanding planet formation

So why is this diversity of exoplanets surprising? Before the first exoplanets were discovered, our models of solar system formation were rather self-centered — we assumed that all solar systems were like ours. Previous models of solar system formation were all finely tuned to ours. We used to think all solar systems around other stars looked like our own, with small rocky planets near the center and gas giants farther away . . . until the first exoplanets were found. We then realized that not all solar system structures mimic ours and that there is actually an enormous diversity of planetary mass, composition, order, and organization. One of the major realizations was that other stars, not only our Sun, have planets and star systems of their own, including red dwarfs, supergiants, and other objects.

TIP

And on that subject, although most exoplanets orbit a star, some appear to be completely unbound gravitationally to a host star. These exoplanets are called *rogue planets*, meandering through space without a fixed orbit.

Exoplanets Come in Many Shapes and Sizes

Although people (teenagers in particular) tend to hate being labeled, categorization comes in handy when you're looking at exoplanets. After examining enough exoplanets to know that many share the same characteristics, scientists have loosely grouped them into several main types (see Figure 7-2): terrestrial, gas giants, Neptune-like, and super-Earth exoplanets. There are also some that don't fit into any of these groups and defy characterization (again, like teenagers).

WATER WORLDS?

Recent studies of exoplanet mass and size led scientists to hypothesize that some exoplanets have lower densities than Earth. In the constellation Lyra, for example, there are two extrasolar planets, Kepler-138d and Kepler-138c, whose densities suggest that the planets could consist of more than 50 percent ocean water. Although our Earth's surface area is more than 70 percent water, that water accounts for less than .05 percent of Earth's mass. Finding exoplanets with potentially a much higher amount of liquid water would suggest entirely new potential in the search for life in the universe — although, because the density of liquid water and frozen ice is pretty similar, these worlds could just be balls of ice.

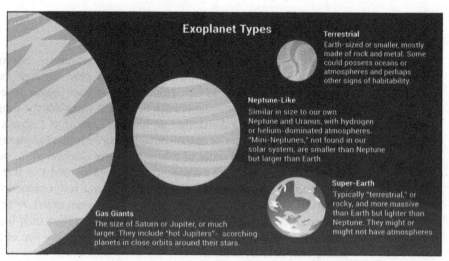

FIGURE 7-2:
Types of
exoplanets.

Exoplanet properties (mileage may vary)

Within our solar system, not all planets are created equal. Some are solid (consisting of rock, minerals, and metals,) with or without a gaseous atmosphere. Earth, Mars, Venus, and Mercury are all examples of rocky planets. Others, like Jupiter, Neptune, Saturn, and Uranus, are the "gas giants" composed of hydrogen, helium, and other gasses, without a solid surface you could walk on. Exoplanet material composition is similar; the ones scientists have been able to study so far seem to be composed of rock, minerals, gasses, and water (frozen or liquid), but the proportions may be different from that of the planets in our solar system.

What else do we know about exoplanets? Gas giant exoplanets likely have the same kind of rings that orbit our solar system's gas giants. Scientists can calculate properties such as mass and diameter for many exoplanets to determine similarity or difference from planets within our own solar system. Most of the exoplanets studied to date range in size from the size of Mercury to 2.5 times the size of Jupiter, with masses varying similarly.

It's always cool to be a part of the "in crowd," right? Well, let's think this one through. The Sun is hot — about 10,000°F (5500°C). The other planets in our solar system range from Venus at 867°F (464°C) to Neptune at −320°F (−196°C). Extrasolar planets, those outside of our solar system, have an equally vast temperature span, ranging from 7800°F or 4300°C (KELT-9b, a massive gas giant discovered in 2017) down to ones that are freezing cold. An example of a colder exoplanet, and one with a distinctly hard-to-remember name, is OGLE-2005-BLG-390L b; this world is a chilly −370°F (−223°C), only 50 degrees above absolute zero! (No wonder it's nicknamed "Hoth.")

What's in a name?

Planets in our solar system have simple, familiar names — Neptune, for example. If you pick up an astronomical catalog and read through exoplanet names, though, they don't exactly flow from the tip of your tongue. There are a few different ways in which exoplanets can be named. Consider the exoplanet orbiting a star in a well-known constellation, like the case of 51 Pegasus b, an exoplanet that orbits star number 51 in the constellation Pegasus. Exoplanets are then given letters in alphabetical order of when they were discovered. In our example here, 51 Pegasus b is the first exoplanet discovered around the star 51 Pegasus. (By convention, there is no "a" planet because that letter sometimes refers to the star itself.) If the star has its own name, that's even easier — Fomalhaut b is the first exoplanet orbiting the star Fomalhaut. Some stars only have catalog numbers, and so HD 40307 b is the first exoplanet orbiting the 40307th star in the Henry Draper (HD) catalog.

Some exoplanets are also named after the telescope or spacecraft that discovered them. The planet-hunting Kepler mission, for example, has thousands of exoplanets named after it — a world like the terrestrial exoplanet Kepler-186f represents the 186th planet-hosting star found by the Kepler space telescope, and is the fifth planet candidate around that star.

TIP

The International Astronomical Union (IAU) is in charge of naming objects in our solar system and beyond, and they have run public contests to identify potential names for exoplanets. Check out `https://www.nameexoworlds.iau.org/` for more information on past versions of the contest, and to prepare for the next round — who knows, maybe your name could be chosen! A 14-year-old suggested the name "Aegir" (after the Norse God of the sea), and it's now the official name of Epsilon Eridani b!

(Non) Flaming giant balls of gas and Neptunian exoplanets

The largest planets in our solar system are Jupiter (more than ten times the diameter of Earth) and Saturn (about nine times the diameter of Earth). They are, together with Neptune and Uranus, called the "gas giants" of our solar system because although they may have cores of metal or rock, most of their mass is in the form of compressed gas. Really compressed gas — Jupiter is more than 300 times the mass of Earth! Exoplanets that resemble this structure are called *gas giant exoplanets*. A subgrouping within these gas giants are the *Neptunian exoplanets*, so called because they're about the size of Neptune (four times the diameter and 17 times the mass of Earth). They are smaller than the full-blown gas giant exoplanets but still have atmospheres predominantly of helium and hydrogen gas. There's even a further subgrouping, the *mini-Neptune exoplanets*, which have similar compositions but are smaller than Neptune — still larger than Earth, though!

The gas giants are all in the outer reaches of our solar system, leaving the smaller planets closer to the Sun. Scientists assumed this would be the case for exoplanets as well, but nature proved us wrong; in fact, the first exoplanets we found are what are now called *hot Jupiters* — huge gas giants that orbit much closer to their star than any planet in our solar system. These hot Jupiters receive vast amounts of energy from their stars, resulting in surface temperatures of thousands of degrees. Some may also have thick clouds in their atmospheres.

They really are super! Super-Earth exoplanets

Superb (better than good) and superfluous (more than needed) are but two of the words we use that come from the Latin *super*, meaning *beyond, above,* or *over the top*. Super-Earth exoplanets got their name because they're just like us, but better (did we really just say that?). Not so much better, really, but they are larger than Earth. Super-Earths are defined as being smaller than Neptune but larger than Earth — anywhere from 2 to 10 times the mass. They can be rocky, gassy, or both; we have nothing to compare planets of this size and composition to in our solar system.

So far, astronomers have discovered about 1600 super-Earth-sized exoplanets. Scientists aren't sure what the minimum size is for a planet to transition from a rocky world into a ball of mostly gas without a solid surface. In 2016, observations made with the Spitzer Space Telescope revealed a super-Earth world with huge temperature swings from side to side. Astronomers think that the reason for these fluctuations is that this world, 55 Cancri e, is so close to its star that it's tidally locked, meaning one side always faces the star 55 Cancri and has eternal sunshine, whereas the other side always faces away. This world orbits and rotates every 18 hours. Observations from Spitzer showed that the hot side is about 4400°F (2700 K) and the cold side is about 2060°F (1400 K). This is a huge temperature difference — and what's more, planetary scientists believe that the hot side is covered with liquid lava flows!

Look no further than Mother Earth for terrestrial exoplanets

All these planets are great, but what about finding another Earth? A terrestrial exoplanet is defined as any rocky, Earthlike body found outside of our solar system. These exoplanets are up to twice the size of the Earth or smaller (anything larger and they'd fall into the Super-Earth category) and are composed of materials similar to Earth such as metal and rock. They may have oceans, deserts, atmospheres, and other conditions that would make life possible. So far, we've found about 200 terrestrial exoplanets, with more and more to come — these smaller worlds are harder to detect than bigger ones, so likely we've only scratched the surface in our current detections.

These rocky worlds have solid surfaces, likely made of materials similar to Earth. They may have metallic cores and atmospheres, but those aren't required; the main requirement to be a terrestrial exoplanet is size. Astronomers estimate that anywhere from 2 to 12 percent of stars in the sky could have rocky planets at the right distance to potentially support liquid water at their surfaces, which could mean 300 million of these worlds in our Milky Way Galaxy alone!

As they detect more and more terrestrial exoplanets, astronomers have noted a lack of planets with sizes about 1.5 to 2 times the diameter of the Earth. It's possible that we've just missed them, but it's also possible that there's a breaking point between the rocky terrestrial planets and some process that causes larger worlds to quickly increase in size. More observations will help explain this potential mystery.

Looking Under (or Around) Hidden Rocks: Exoplanet Detection

Why do we look for exoplanets? The short answer is, it's human nature. Since the earliest of times, humans have set their sights on exploration. Ancient Egyptians traveled to Africa in the 4th century BCE; the Greeks and Romans traveled extensively in the 1st centuries BCE and CE in the search for new lands. Voyages of exploration and discovery have always been part of human civilization, and using the stars as navigation tools is literally a tale as old as time.

Those of us who gravitate (ha!) to astronomy and physics have an innate desire to understand what we see in the night sky. Are we alone out there? Where else might life exist? Combine our fundamental drive to explore with a keen sense of urgency around mapping the cosmos, and presto — you're an exoplanet detective. Thanks to dedicated space missions like the Kepler space telescope, we now suspect that there are many more planets than stars in our galaxy. There could be more than a trillion planets in our own Milky Way Galaxy alone!

Observational techniques and biases

Although you can look out your window and see our Sun and Moon, exoplanets are considerably harder to see. Most exoplanet observations are therefore indirect — they can be observed through their effects on stars and other bodies. Regardless of the technique, it's easier to see exoplanets that stand out in some way. In some cases, the bigger the planet compared to its star, the easier it is to see (so we can find an Earth orbiting a tiny star more easily than an Earth orbiting a massive star). Some methods allow scientists only to find planets whose orbits happen to be exactly edge-on as seen from Earth, and planets orbiting stars that are closer to Earth are often easier to find than ones orbiting more distant stars.

For all these reasons, our current catalog of exoplanets is biased towards higher-mass exoplanets that are closer to their stars, stars that are closer to Earth and/or dimmer, and/or ones whose orbits have a particular orientation. The conclusion? Our exoplanet catalog is far from complete. When we look at the formation and distribution of exoplanets, we have to take these biases into account. Just because we have mostly found larger exoplanets, for example, this doesn't mean that most exoplanets are large — it just means that smaller exoplanets are harder to detect.

Finding the distance between two points with radial velocity

Watch a marble running down a track and notice that the marble spins and wobbles slightly as it follows the curves of the track. As an exoplanet orbits its host star, the star moves in an elliptical motion in response to the exoplanet's exerted gravity. These movements of the host star change its light spectrum, exhibiting redshift (moving toward longer wavelengths) or blueshift, moving toward the shorter wavelengths. Tracking the motions of an exoplanet via these shifts is known as the *radial velocity method*, as shown in Figure 7-3.

Radial velocity was the first method used to find exoplanets. The advantage of radial velocity measurements is that they can be performed using specialized instruments to measure spectra on roughly 1-meter diameter mirror and larger Earth-based telescopes, but the disadvantage is that consistent observations must be made month after month for years in order to detect these faint wobbles. The longer the time series of observations, the better resolution in terms of radial velocity.

One of the most famous radial velocity exoplanet discoveries was the 1995 detection of a planet orbiting the star 51 Pegasus, with a minimum mass of about half that of Jupiter. This discovery really launched the exoplanet detection era, and it's only accelerated from there. About 20 percent of our current known exoplanets were first detected using the radial velocity method. One downside, in addition to the long time period required for observation, is that although the period of a potential exoplanet can be measured to relatively high accuracy, the mass depends on the orbit's tilt . . . which we can't see. The more the orbit is tilted, the more mass can be hidden as it tugs in directions not radial to us.

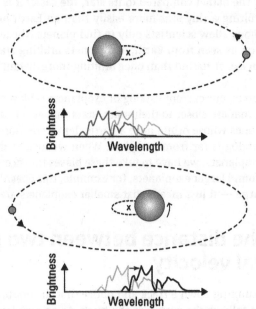

FIGURE 7-3:
Radial velocity.

To cut through the confusion, astronomers talk about a "minimum" mass:

$$M_{min} = M_{true} \sin i$$

M_{min} is the minimum mass, M_{true} is the actual mass, and i is the orbit's inclination or tilt. The real mass could actually be higher depending on the value of the inclination of the orbit. Radial velocity measurements can be combined with measurements using other techniques, though, to help reduce this uncertainty.

Transiting, TTV, and gravitational lensing

Transiting is the act of moving something from one place to another. Public transportation lets you take a bus or train from Point A to Point B, for example. Transiting, in the realm of astronomy and physics, refers to one celestial body passing across another body. When an exoplanet transits across its host star, the host dims during this passage, as shown in Figure 7-4. By making very careful measurements of the brightness of the star as the exoplanet transits across it, astronomers can estimate the diameter of an exoplanet as well as the inclination of the plane of the orbit through a technique called photometry. That's already one big advantage over the radial velocity technique, but there's more! If astronomers are able to watch a transit repeating, they can determine the orbital period of the transiting planet.

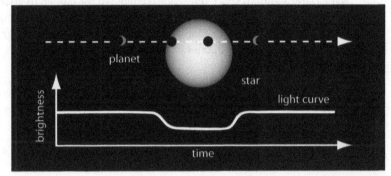

Courtesy of NASA/JPL-Caltech / Public Domain

FIGURE 7-4:
Transit diagram showing a dip in brightness.

The transit method works best on larger planets that are relatively close to their host star; those planets will block more light and will transit more frequently. Another version of the technique is called *transit timing variations*, and this technique involves making a prediction of when a known transit should occur. If the transit repeatedly occurs sooner or later than expected by different amounts, there's evidence that other planets could be in the system, or even a potential exomoon orbiting the exoplanet.

The Kepler space mission was designed to detect exoplanets using the transit technique. Kepler initially stared at a single portion of the sky near the constellation Cygnus for four years and a set of fields around the ecliptic during an extended K2 mission for another four years. It found more than 2600 transiting exoplanets, including many systems with planets orbiting binary stars. A follow-up mission to Kepler, the TESS (Transiting Exoplanet Survey Satellite) mission, launched in 2018, is able to search for exoplanet transits across much of the sky. TESS is focused on finding smaller planets transiting nearby, bright stars.

Another exoplanet detection mission is ESA's CHEOPS (Characterizing ExOPlanet Satellite.) It's made follow-up observations of known exoplanets but has also detected new worlds using the transit method. All these space telescope observations require careful follow-up from ground-based telescopes to confirm that candidate exoplanets are actually exoplanets and not false detections.

One other important method of detecting exoplanets is called *gravitational micro-lensing*. This technique involves looking for the effect of gravitational bending of light by a star when another star passes in front of it. The gravity of the star in front will briefly increase the brightness of the background star's light. If the background star has a planet orbiting it, that brightness spike will have two peaks. This technique has been demonstrated by ground-based telescopes, and the future Nancy Grace Roman space telescope (scheduled for launch in 2026) will use gravitational microlensing to search for exoplanets near the center of the Milky Way Galaxy. See Chapter 12 for more about how astrophysicists use gravitational lensing to study stars and galaxies.

Ignore the garbage, but know your debris disk

Putting your exoplanet detection cap back on, imagine you're a main-sequence star floating through life.

REMEMBER

Go to Chapter 5 for the details on the H–R diagram; stars along the main sequence are in the phase of life where they generally burn hydrogen into helium. In many cases, remnants of past asteroid collisions form loosely gravitationally-bound rings that orbit these stars. These circles of cosmic debris and dust are called *debris disks*, and they can be very helpful in studying exoplanets. The first debris disk was discovered in 1984; a huge disk around the star Beta Pictoris was found using a ground-based telescope with a 2.5-meter diameter mirror.

Sometimes called planet detectors, debris disks usually have a large percentage of big dust grains. How big is big? Not very large, actually — these grains are up to a millimeter in diameter, and they give off thermal signatures which can be detected using radio telescopes like ALMA (Atacama Large Millimeter/ submillimeter Array) in Chile. Collisions and gravity in the debris disk surrounding a star affect these dust particles as they spread into a relatively coherent disk, unless there is a planet. A planet in the debris disk creates a chaotic zone where the dust no longer is stable, and eventually produces a gap in the disk that looks like a ring. The width of the gap is proportional to the mass of the exoplanet, so measurements of gaps can help reveal a key exoplanet parameter.

Take a picture! The future of exoplanet detection

All these indirect methods are clever, but ultimately a bit unsatisfying. Scientists really want to see an exoplanet and image it directly. Because an exoplanet's light is usually billions of times dimmer than its star, its light needs to be blocked with a physical cover called a starshade or coronagraph to have any hope of imaging it.

These observations also need to focus on planets around nearby stars because those will be brighter and are usually taken using a space telescope to avoid blurring from Earth's atmosphere. Figure 7-5 shows the first ground-based image of a multi-planet exoplanet system around a sunlike star. This observation used the SPHERE instrument on the European Southern Observatory's Very Large Telescope — and yes, it is indeed very large, made up of four separate telescopes with mirrors 27 feet (8.2 meters) across. TYC 8998-760-1, a young star similar to our Sun, is visible in the top-left. By blocking out most of the starlight with a coronagraph, SPHERE revealed the two giant planets at the middle and bottom right of the image.

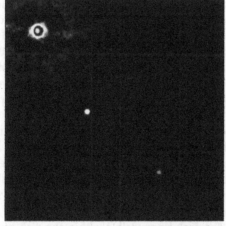

FIGURE 7-5:
Two giant exoplanets orbiting the star TYC 8998-760-1.

Courtesy of ESO/Bohn et al.

Even better direct observations of exoplanets will be coming soon. The upcoming Nancy Grace Roman Space Telescope (planned for launch in 2026) will have a coronagraph instrument specifically designed to look for exoplanets at both visible and infrared wavelengths. It will be able to image planets down to Jupiter-sized ones orbiting stars like our Sun, and will also be able to perform spectroscopy on their atmospheres to determine their composition. The James Webb Space Telescope (JWST) has also begun taking direct images of exoplanets.

The Nitty-Gritty of Exoplanet Formation

With respect to the "clouds of dust and gas" discussion about stars forming in protostellar nebulae (see Chapter 5), scientists are lucky that stars aren't greedy. Burgeoning stars don't use up all the gas and dust around them; stars leave dust behind, and that dust can be rich in materials like silicon, carbon, and even iron.

Bits of leftover material ejected from a forming star can stick together; these blobs start sticking to others, and eventually start expanding into the perfect source material for the development of a young planet.

Planetary birthrights: Protoplanetary disks

As the process of star formation completes, the leftover material from the protostellar nebula collapses down into a disk surrounding the newly formed star; this disk is called a protoplanetary disk. But why does the material collapse?

TECHNICAL
STUFF

This collapse takes place because something shocked an otherwise stable cloud and caused it to compress just enough that gravity took over. A local portion of the protostellar nebula collapses down under the force of gravity to make a star, and that initial shove is never perfectly centered, so the material in the collapsing nebula ends up rotating like a wad of paper hit on the edge. This rotating mass evolves into a system with a rotating disk and a central bulge where the star continues to form. This action is similar to what would happen if you just keep spinning pizza dough tossed into the air — you'd get one very flat pizza.

Orbital motion produces a centrifugal acceleration that can cancel out the star's gravitational pull in the equatorial direction only. Any material above the equatorial plane of the star collapses down onto the forming pizza — er — forming disk. Figure 7-6 shows an image of the protoplanetary disk surrounding the young star HL Tauri, about 450 light-years from Earth in the constellation Taurus. The image was taken with the ALMA telescope array in Chile, observing at wavelengths of around a millimeter. The dark rings visible within the disk could be locations of planets that are forming.

FIGURE 7-6:
Protoplanetary disk surrounding the young star HL Tauri.

Courtesy of ALMA (ESO/NAOJ/NRAO)

Eventually, this leftover material continues to swirl around the star in a proto-planetary disk. Clumps of this disk stick together through a process operating under gravity called *accretion*.

TIP

This early formative object is called a *planetesimal*, and it continues to grow and acquire material. Because planets form out of this disk made from leftover material, the chemical composition of the original cloud determines the composition of the eventual planets.

TECHNICAL STUFF

Eventually, if enough material is gathered up, the planet can heat and differentiate (heavier materials like metals sinking to the core, and lighter materials like rock rising to the surface). Smaller bodies can remain partially undifferentiated. After enough material sticks together and becomes circular under the influence of gravity, with heavy material on the inside and lighter material on the outside, a planet is born.

Planetary composition depends not only on the composition of the protoplanetary nebula (go to to Chapter 14 for more about different populations of stars), but also on where in the protoplanetary nebula that formation occurred.

Jupiter is thought to have formed just past what's sometimes called the *snow line*, or the distance from the Sun where temperatures cooled just enough that volatiles like hydrogen and helium could condense and be captured. The ice giants of Uranus and Neptune formed farther out, at a carbon-monoxide line where carbon-rich ices formed. They have higher concentrations of water ice, as well as compounds like ammonia and nitrogen. Finally, at the end of the planet formation process, any remaining gas and dust is either incorporated into the planets; clumps together into small leftover pieces that become asteroids, comets, and minor planets; or is blown out of the solar system by the solar wind. The protoplanetary nebula is gone, the dust clears, and you're left with a brand-new baby solar system.

Get moving! The dynamics of exoplanet systems

As astronomers began detecting exoplanet systems, one of the biggest surprises was how different these systems are from our own solar system. Scientists had to completely update their models of how exoplanet systems form and evolve, and new information is coming in all the time as we find more odd but wonderful configurations. A key area of interest for astronomers here is the dynamics of exoplanet systems.

Observations of protoplanetary disks, for example (see Figure 7-7), reveal gaps that could be where planets are in the process of forming. These disks could also have clues as to whether exoplanets migrated after formation, or whether they formed in place and have stayed there. Another data point is the configuration of exoplanet systems themselves. We've found systems with planets located very close to their stars, even closer than Mercury is to our Sun. Astronomers have found planetary systems with giant planets in the inner solar system, as well as systems where planets are very close to each other — much closer than any planets in our own solar system. Star types can be different from our Type G Sun (see Chapter 5 for more about stellar types), with many exoplanets orbiting Type M red dwarf stars that are much cooler than our Sun.

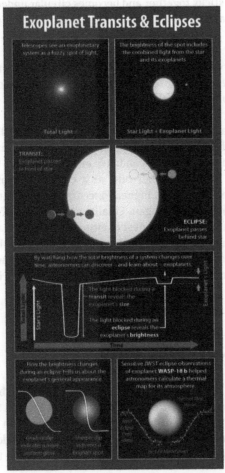

FIGURE 7-7:
How JWST observes exoplanet atmospheres during eclipses and transits.

Courtesy of NASA/JPL-Caltech (R. Hurt/IPAC) / Public Domain

One of the more puzzling types of exoplanets that have been found are so-called "hHot Jupiters," or gas giants that are located very close to their star. These famous, or perhaps infamous, planet types were some of the first exoplanets discovered. They are extremely puzzling because although they seem to have the same size and composition as gas giants like Jupiter, they are so close to their stars that their surface temperatures are likely in the thousands of degrees. 51 Pegasus b, for example, is a hHot Jupiter. But our whole theory about the snow line would suggest that gas giants can't form that close to their stars, forcing scientists to start exploring options. One possibility here is that they could have taken the scenic route, forming in the outer solar system and then migrating inwards.

All these observations feed into computer models that astrophysicists and planetary scientists have been developing and refining to explain how planetary systems form and change over time. Models take into account the initial conditions thought to be in a protoplanetary disk, and then these models may be run over a long period of time. All the while, scientists apply the laws of physics and dynamics (including gravity and magnetic fields,) and also take into account the random variations occurring in a very complex system over time. These models are run repeatedly, giving a different range of outcomes each time, and those outcomes are compared with our actual observations of planetary systems.

What the Hail . . . Exoplanet Atmospheres

No, smog isn't an atmosphere in and of itself, though if you live in Southern California, it can sure feel like it!

TIP

Atmospheres are the various layers of gases that envelop any celestial body. Our atmosphere on Earth is a complex system consisting of multiple layers, starting with the Earth's surface and expanding out. Do exoplanets have atmospheres? They sure do, and they're absolutely critical to our ability to observe and gather data on exoplanets.

REMEMBER

One way in which exoplanet atmospheres are measured is through spectroscopy, or breaking light down into its constituent wavelengths as it interacts with matter. Exoplanet atmospheres can be resolved through the process of transiting. You already learned how transiting is the passing of an exoplanet in front of a star such that the amount of light it blocks is detectable (refer to Figure 7-4). How does this help us learn about an exoplanet's atmosphere? The act of transiting allows starlight to pass through the exoplanet's atmosphere; using techniques such as high dispersion spectroscopy, astronomers can identify spectral lines that appear during a planetary transit (refer to Figure 7-7). By comparing the star's spectrum before and after an exoplanet transit, new spectral lines that appear only during the transit can be identified as coming from atoms and molecules in the exoplanet's atmosphere. Similar techniques can also be used during exoplanet eclipses, when an exoplanet moves completely behind its star.

It's keenly important to be able to separate an exoplanet's spectral signature from that of its host star, and the James Webb Space Telescope (JWST) helps scientists to do just that. Launched in 2021, JWST has many capabilities including specialized instruments to probe exoplanet atmospheres and return data on their composition. For example, JWST has provided evidence that the gas giant exoplanet WASP-39 b has sulfur dioxide and carbon dioxide in its atmosphere, and has even found evidence of water vapor in the atmosphere of exoplanet WASP-18b, in the hot Jupiter category.

The future ESA Ariel mission, planned for launch in 2029, will be devoted to studying about 1000 known exoplanets in detail. It will use spectroscopy to search for signs of various chemicals in the atmospheres of these exoplanets, including carbon dioxide, methane, and water vapor, and will look for clouds and atmospheric variations.

Can Life Be Found on Exoplanets?

The search for life in the universe takes us to unexpected places. Humans have studied the far reaches of our own planet, and have found life in all sorts of strange locations — deep at the bottom of our oceans, buried within ice sheets, and in the highest and driest deserts of Earth. The search for life in our solar system beyond the Earth has focused on Mars, which could have had a warm, wet climate a few billion years ago, and the ocean moons of the outer solar system, such as Jupiter's moon Europa, which could have the necessary conditions to support life today.

One of the most amazing discoveries of the past 25 years is that exoplanet systems seem ubiquitous — from thinking that planets could be rare occurrences to the current understanding that when you look up at the sky, almost every star you see has a planet around it! The big implication here is that there could be more and more possible places for life to exist.

Despite what Hollywood would have us believe, packing up and moving to another planet in our solar system isn't a simple proposition. Living on an exoplanet? Even less conceivable, given our current knowledge of exoplanets and known requirements for human life (and let's not talk about the fact that we've yet to invent a warp drive). But what about finding any life at all in a galaxy (or solar system) far, far away? Are any exoplanets habitable, and could they actually be inhabited?

Goldilocks had it right: The "habitable zone," or conditions for life as we know it

Although you may think you need phones, laptops, and the Internet in your life, the requirements for human life are basic: We need air, food, water, heat, and shelter. What about a planet's ability to support life?

TIP

A planet's habitability requirement is a bit more complex in that scientists must think about every element needed to make a planet able to support life. But — and this is a big caveat — research must also recognize the limitation in those thoughts. Current ideas of habitability are based on human life and our solar system because that's the environment we humans inhabit. Were life to take on another form completely, these notions would have to evolve along with it.

That said, the main requirements for life (at least, life as we know it) are

>> **Liquid water:** All life on Earth is based on water, and it's considered a universal solvent.

>> **The right chemical elements:** These are materials like carbon, hydrogen, nitrogen, oxygen, phosphorus, and sulfur, and they make up a large percentage of organic material.

>> **An energy source:** To drive the chemical reactions that sustain life.

Because the chemical elements listed above are fairly common (and it's hard to test for energy sources from afar), most scientists use the presence of liquid water as a good starting point in the search for life beyond Earth. They've defined what's called a *habitable zone*, the orbital distance from a star to where water could be liquid at the surface of a planet. In our own solar system, the Earth is in the habitable zone — as it should be, because it's clearly inhabited. Venus is a bit too close to the Sun, and its thick atmosphere makes the surface lead-meltingly hot. Mars is a bit too far from the Sun — its thin atmosphere makes the surface far too cold for liquid water. You can see why the habitable zone is sometimes called the "Goldilocks zone" — not too hot, not too cold, but just right!

If you are looking for life in other planetary systems, you'll be particularly interested in those that have terrestrial or rocky planets in their habitable zones. Because the zone depends on temperature, the size and distance depend on how hot and bright the star is, as shown in Figure 7-8. A type M star, one with a cooler, smaller red dwarf, has a small habitable zone that's much closer to the star. A hot massive type A star will have a larger habitable zone that's further out.

Habitable zone size

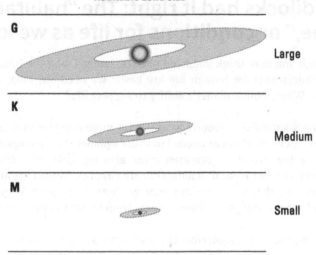

Courtesy of NASA/JPL-Caltech / Public Domain

FIGURE 7-8:
The size of a star's habitable zone depends on the star's brightness.

Earth obviously is in a habitable zone, but are any other planets currently taking reservations? Yes! So far, our various search methods have found dozens of exoplanets that seem to be in their star's habitable zone. For example, the exoplanet closest to Earth is Proxima Centauri b. At a mere 4 light-years away from Earth, it's thought that Proxima Centauri b is solid and rocky. It orbits close enough to its parent star (Proxima Centauri) that it's within the habitable zone, and this indicator opens the door to the existence of life there. Another star system, the TRAPPIST-1 system, is about 40 light-years away from Earth and has not just one, but seven rocky planets orbiting a cool Type M red dwarf star in the habitable zone.

Twinning: Why haven't we found Earth 2.0?

What's up with all these Earthlike planets? Has anyone actually found another world like Earth? Sadly, the current answer is no. Happily, though, this discovery could be getting closer! A planet like Earth, orbiting a star like the Sun, and in the habitable zone is the prize that all planet-hunters are seeking. There are some near misses, but still no second Earth — at least not yet.

For example, Kepler-452b (see Figure 7-9) was discovered in 2015 as an exoplanet about 1400 light-years from Earth. It's close to Earth's size (roughly 60 percent larger in diameter) with a 385-day orbit — very close to Earth's 365-day orbital period. Its similar mass suggests that Kepler-452b could be a rocky planet like Earth, and its parent star is eerily similar in mass and luminosity to our Sun. One difference is age — just a number, to be sure — but in this case, Kepler-452b's parent star is about 1.5 billion years older than our 4.5-billion-year-old Sun. Although Kepler-452b seems pretty similar to the Earth, its larger size suggests that if it's rocky, its mass is about five times that of Earth; its size places Kepler-452b in the super-Earth category — not Earth's twin, at all, but getting closer. Figure 7-9 shows how the Kepler-452 system compares to our own solar system.

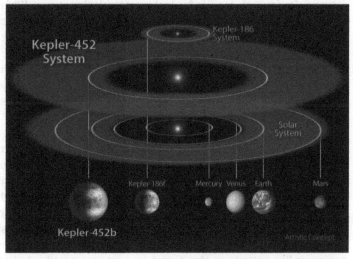

FIGURE 7-9: Comparison of Kepler-452b to our solar system.

Courtesy of NASA/JPL-Caltech/R. Hurt / Public Domain

Other potential Earthlike exoplanets have similar issues. Gliese 667 Cc seemed like a good candidate, but it turned out to be about 4.5 times as massive as the Earth. It orbits a cool red dwarf star with an orbital period of 28 days, making it quite close to its star but technically within the habitable zone because the star is so much cooler than our Sun. Not exactly Earthlike, but fairly close! One disadvantage, though, is that at that orbital distance, the planet may be subject to dangerous solar flares from its star. Another planet orbiting a red dwarf star, Kepler-186f is one of the closest we've found to an Earth-sized body at only about 10 percent larger than the Earth. Its star, Kepler-186, is so faint that the planet orbits at the outer edge of its habitable zone and only receives about one-third of the energy that the Earth receives from our Sun.

Another cool possibility for a habitable world could be a large moon of a giant planet. In our own solar system, the habitable zone ends before you get to Mars. Jupiter's moon Europa is about the size of Earth's moon and is covered with a layer of ice on top of an ocean of liquid water. Scientists think this ocean contains more water than all of Earth's oceans combined; NASA's Europa Clipper spacecraft will work to see if Europa could be habitable. It's possible that exomoons orbiting exoplanets could also be habitable. Although we haven't found any exomoons yet, scientists are still looking.

Smells like life: Searching for biosignatures in exoplanet atmospheres

The next step after deciding if an exoplanet is habitable is to see if it's inhabited. Astronomers search for life by looking for biosignatures. When you write your name at the bottom of a driver's license application or other important piece of paperwork (or digital file), you're adding your signature to the document. This is your unique method of identification, one that shows your approval of the document in question and verifies your identity. A biosignature or biomarker is similar but on the physical level; any substance (isotopes, elements, or other characteristics) that provides evidence of life, past or present, is called a *biosignature*.

For exampe, carbon dioxide is produced by the human body during metabolism. It's exhaled whenever you breathe out and as such is a biosignature of human life. The same is true for the oxygen produced by plants via photosynthesis. Either of these elements would be an important biosignature in the search for life on exoplanets because they're both indicative of the ability to support life, human or otherwise. Gases like oxygen and methane are unlikely to persist unless they are constantly being replenished. Although they can be emitted by non-biological processes like volcanoes, significant amounts are possible indicators of life.

One way to look for biosignatures is to use the same methods used to study exoplanet atmospheres but with a search for these biosignature gasses. One particularly interesting search combination could be oxygen and methane. Although these gases normally react with each other, the presence of some form of life could produce a chemical disequilibrium that would allow their molecules to coexist.

The Drake Equation and the search for intelligent life

Biosignatures are absolutely useful in the search for life, but a major goal of planet-hunters is to find intelligent life. If microbes are hard enough to find, how much harder would it be to detect an entire civilization? As it turns out, it would be easier because a civilization might give off its own technological signals — a technosignature — and could even try to talk to us directly.

Back in 1961, American astrophysicist Frank Drake (he later went on to help found the SETI Institute, which stands for "Search for Extraterrestrial Intelligence") came up with an elegant way to describe all the different variables that go into the search for intelligent life. This description became known as the Drake Equation:

$$N = R_* f_p n_e f_L f_i f_c L$$

The goal of the Drake Equation is to derive a way of calculating N, the number of potential civilizations in our Milky Way Galaxy that are giving off detectable electromagnetic signals. The factors that determine it are

>> R_* = the rate at which stars suitable to support planets that contain life form, per year

>> f_p = the fraction of those stars with planets

>> n_e = the number of planets per solar system that are in the habitable zone

>> f_L = the fraction of planets which are habitable that actually form life

>> f_i = the fraction of life-bearing planets that form intelligent life

>> f_c = the fraction of intelligent-life-bearing planets that have a civilization that produces detectable signals

>> L = the number of years that such a civilization produces signals that can be detected

When all these factors are multiplied together, they theoretically produce an estimate of how many civilizations could be giving off detectable signals related to extraterrestrial intelligence. The SETI Institute and other groups have been searching for these signals, at first with giant radio telescopes and more recently using optical and other methods, but so far without success. Scientists have had better luck in understanding the values of the parameters that would go into the Drake Equation. For example, we now know that F_p, the fraction of stars with planets, is very close to 1 — and that N_e, the number of planets per solar system in the habitable zone, is likely sizeable as well.

One final puzzle, then, is where are the aliens? If planets are as common as they now seem to be, why haven't aliens come calling yet or at least sent us a signal? UFO sightings aside, this puzzle is sometimes referred to as the Fermi Paradox after physicist Enrico Fermi. Maybe Earthlike planets are much less common than they seem, or maybe there's another factor at play here. Some scientists have suggested that perhaps L in the Drake Equation, the amount of time that an intelligent civilization gives off detectable signals, is very short — meaning that we'd have to time our observations just right to detect them. There's some evidence supporting this theory; over about a 50-year period, humans have transitioned from outward-pointing broadcast television and radio waves (sending detectable signals off into space in all directions) to cable and satellite and streaming telecommunications (more downward-focused and giving off much less stray information). Fifty years is not very long at all in the grand scheme of things so whatever the odds, scientists will certainly keep looking.

Chapter **8**

White Dwarfs, Black Holes, and Neutrinos, Oh My!

"I can jump ten feet in the air!" "No, you can't." "Yes, I can." "Prove it!"

Life is full of declarations; some have merit, but many do not. New ideas, particularly controversial ones, are typically met with skepticism until they are verified and proven. Take John Locke's infamous *tabula rasa*, or "blank slate," argument of 1689, for example. Locke theorized that newborn babies have a completely empty mind, one waiting to be filled from experience alone. This theory disregarded the role of genetics and biological predisposition in forming one's personality and was subsequently disproven, but at the time it stood as one of the foremost theories in social development.

Astrophysics (and most branches of science, for that matter) operates in a similar manner. When you're driving without GPS and have to make a decision about which way to turn, you use knowledge and experience to make your best judgment. Scientists do exactly that; they use the information available to them to create theories, then use data and observation to either prove or disprove those theories.

As 20th-century astronomers developed theories about the stars and beyond, telescope engineering advanced in leaps and bounds towards observing at different wavelengths. It became possible to prove (or disprove) astronomical predictions such as black holes and neutron stars. Telescope and satellite detection technology is constantly improving, and astronomers may soon have an answer on which other theories will bear definitive fruit. This chapter describes particularly interesting astrophysical objects; some were first predicted through theories and others were detected first, requiring theories to be updated (or, in some cases, invented) to explain them.

Snow White and the Seven . . .

To walk in the footprints of a giant! No, stars don't walk and they don't leave footprints, but they do vary widely in their size and luminosity. *Giant stars,* for example, have a large radius compared to their temperature and mass; their large size means that they have a large surface area that gives off electromagnetic radiation and, as such, appears extremely bright. One of the more famous examples of a giant star (and, in fact, the biggest known star in the universe) is UY Scuti. This *hypergiant,* or a star even bigger than a supergiant, is located in our very own Milky Way. UY Scuti is about 9500 light-years from Earth and has a radius 1,500 times as large as our Sun. On the H–R diagram (see Chapter 5 for more information on how to interpret the H–R scale), giant stars occupy the top-right corner, just above the main sequence.

The term *solar radius* is used to describe the size (radius) of a stellar object relative to the Sun. This unit is defined as the radius to the Sun's photosphere where photons are more likely to escape than they are to be scattered or absorbed. At this depth the Sun's radius is about 432,000 miles (695,000 km).

At the other end of the spectrum (Ha! See what we did there?) are *dwarf stars,* ones with relatively lower mass and luminosity. When we classify stars on the H–R diagram, stars on the main sequence (those generally burning hydrogen) are dwarfs or (grouped a bit fainter) subdwarfs. These classifications are based on how much gravity someone on their surface would encounter (as they died from the heat). The smallest dwarf and subdwarf stars include red and brown stars that are named red and brown dwarf stars. Yes, this means red dwarf stars can be classified as subdwarfs. Adding to the confusion, further below the main sequence is a band of stellar remnants — the white dwarf stars we discuss in Chapter 5. White dwarfs are not dwarf stars — they are not even stars! They are just leftover cores of former stars. (Yes, this classification system is a bit of a mess!).

Pure as the driven snow, but without the snow: White dwarfs

A white dwarf star is a type of stellar remnant; it's the end state after a star that started with fewer than eight solar masses has evolved and exhausted its fuel. Because white dwarfs have run out of fuel, they no longer generate heat through nuclear fusion. They start out immensely hot and continue to give off heat after they're formed like a cooling coal from a fire.

A typical white dwarf might have about the mass of our Sun but only the radius of the Earth, making its average density about a million times greater than the Sun!

REMEMBER

In fact, a one-centimeter cube of white dwarf material, if brought to Earth, would weigh more than a car. The incredibly high density of a white dwarf is supported by a quantum mechanical behavior called *electron degeneracy pressure*, and this pressure keeps it from collapsing any further.

TECHNICAL STUFF

Electron degeneracy comes from a quantum mechanics concept called the *Pauli exclusion principle*, named after Austrian physicist Wolfgang Pauli. The main idea is that two electrons with the same spin cannot physically occupy the same volume. In a white dwarf, the density is so high that the electrons aren't associated with specific atoms. Instead, all the electrons act like they are in a single atom, with the electrons arranging themselves kind of like they were in a giant set of energy levels. The motion of electrons in these confined energies creates the electron degeneracy pressure. As a collapsing stellar core reaches a density of about 10^6 kg/m³, electron degeneracy begins and a white dwarf is born.

White dwarfs are also unusual in other ways. The larger the mass of a white dwarf, for example, the smaller its radius and the higher its density (up to a maximum mass of about 1.4 times the mass of our Sun). This maximum mass is called the *Chandrasekhar limit*, named after famous Indian-American physicist Subrahmanyan Chandrasekhar. As the mass of the object approaches 1.4 solar masses, the force of gravity gets stronger and stronger and the material of the star is squeezed into a smaller and smaller volume as supported by the electron degeneracy pressure. At the Chandrasekhar limit of 1.4 times the mass of the Sun, electrons are moving at velocities close to the speed of light! At this mass, gravity is strong enough that it overcomes the electron degeneracy pressure, and the object collapses further into a neutron star, which is supported by neutron degeneracy pressure.

BLACK DWARFS, A THEORETICAL EXTENSION OF WHITE DWARFS

After stars burn through their fuel and begin their end-of-life transitioning, there are a few possible outcomes. Some stars will explode spectacularly into a nova or supernova, whereas less massive stars gravitate (not literally) to the white dwarf side. But when a star becomes a certified white dwarf, what's next for our aging friend? Scientists think that when a white dwarf emits all its energy via radiation, that star will cease to radiate anything — no energy means no light, and it also means no heat. In theory, though, that stone-cold star still has some amount of mass left. With no energy and no way to dissipate that mass, our star would exist as a dark object called a black dwarf. There's no way to prove this as the ultimate end-case for white dwarfs because our universe is too young; it's thought that hundreds of billions of years would be required for any white dwarf to reach this stage.

So cool and red hot: Red dwarfs

Who's the coolest star on the block? Red dwarfs, that's who! Actually, they are the coolest stars on the main sequence line of the H–R diagram (see Chapter 5 for the full story there.) Red dwarfs form directly from a star formation region or a protostellar nebula. When enough gas and dust accrete together under the force of gravity, nuclear fusion takes place by burning hydrogen into helium.

There are more red dwarfs than any other type of star, so they are far from uncommon. They're small (from a bit less than half a solar mass down to less than a tenth of a solar mass) and, as such, are relatively dim and can only be observed via telescope. Because they're small and cool, nuclear reactions in red dwarfs are slow and they take a lot longer to run out of hydrogen fuel than larger stars. This means they can live very long lives — in the trillions of years.

REMEMBER

The surface temperature of a dwarf star aligns with its emitted color. Red dwarfs have a surface temperature of around 2500–3000 Kelvin (4000–4900°F), white dwarfs start at 100,000 K (180,000°F), brown dwarfs vary from 1200 K to 1700 K (1700–2600°F), and blue dwarfs are around 8600 K (15,000°F).

Even smaller than red dwarfs are brown dwarfs. These critters tend to be less than 10 percent of a solar mass, placing them between smaller stars and bigger planets in terms of size. Despite their small girth, brown dwarfs are dense — a typical brown dwarf star is only about 20 percent larger than Jupiter, but its mass range can be from 13 to 80 times the mass of Jupiter. When a star gets up to about 80 times the mass of Jupiter, it's usually big enough to begin the long-term fusion of hydrogen into helium.

TECHNICAL STUFF

Why are brown dwarfs called "failed stars"? To be more accurate, brown dwarfs are sometimes not even considered stars. Their relatively small size prevents them from undergoing hydrogen-helium fusion, typically considered a requirement for a star. So why aren't these objects just big planets? In fact, brown dwarfs are dense enough for tritium or deuterium fusion, and some of the more massive ones can even perform the fusion of lithium. These elements can fuse at slightly lower temperatures, though they are also much less abundant, so fusion ends when the stars run out of them.

Brown dwarfs give off very little light in the visible portion of the spectrum, so they are hard to detect using optical telescopes. They do have a strong signature in the thermal portion of the infrared spectrum (flip to Chapter 2 for a refresher on the EM spectrum), and infrared telescopes have been able to detect thousands of brown dwarfs. They likely form directly from a protostellar nebula, but just accumulate a small amount of mass. Gravitational collapse provides less energy, and the core never heats up enough for fusion to really get going before the gas pressure balances gravity (like with Jupiter) and the density stops increasing as the star stops collapsing.

TIP

If brown dwarfs don't really create energy through hydrogen fusion into helium (though small amounts are possible from tritium, deuterium and lithium fusion), how can you even see them? Brown dwarfs radiate out the heat energy from their formation, and this energy that can be seen in infrared observations.

Giant stars

Remember how we said the naming of dwarf stars was kind of a mess? Let's continue the confusion and meet all the giant stars. Looking at the main sequence, astronomers sometimes call those stars that have eight times as much mass as the Sun, or more, giant stars. These are the stars that will die as something more interesting than a white dwarf. Technically, however, they are still classified as dwarf stars, and called things like O and B stars after their spectral type. Stars that are actually classified as Giants (from looking at the gravity at their surface), are stars that evolved off the main sequence and grew larger with lower density when they stopped fusing hydrogen in their core, and moved on to generating energy in other ways. Some giant stars are one solar mass! Those stars we nicknamed giants — the stars with more than eight solar masses on the main sequence — never actually become giants . . . they instead evolve into hyper giants along the top of the H–R diagram! Confusing? Yes.

Because massive stars burn up their fuel quickly, they tend to have shorter, more dramatic lives than smaller stars. Why not go out in a blaze of glory if you can? Although our Sun has been around for 4.5 billion years and probably will last another 5 billion, that 8-solar-mass star will probably burn out in 100 million years, and a giant star that's 10 to 15 solar masses will only last 10 to 20 million years.

Evolved Giant Stars

There are several classes of evolved giant stars, including

>> **Supergiants:** Both large and luminous, these form when a star eight to twelve times the size of the sun departs the main sequence and burns helium.

>> **Subgiants:** Smaller and dimmer than supergiants, these stars are up to eight solar masses and have stopped fusing hydrogen in their core but not yet started burning a shell of hydrogen around the core.

>> **Red giants:** Extremely luminous, these stars form when subgiant stars start burning hydrogen around the core.

Sun-sized stars become both subgiants and then red giants before they exhale their atmosphere as a planetary nebula and leave behind a white dwarf. (Curious about the ultimate end of our solar system? See Chapter 16.) Supergiants are similar, but form when a much larger star transitions from burning hydrogen to helium. They then move on through nucleosynthesis to produce an iron core that collapses into a supernova.

TIP

You can easily see the bright red supergiant star Betelgeuse in the constellation Orion. Another great target for your unaided eye is the yellow hypergiant variable star Rho Cassiopeiae, located in the constellation Cassiopeia.

There Is No Escape: Black Holes

Gravity.

Let's say it again: Gravity. That's almost all you need to understand black holes, but we'll go ahead and dive into the details. You can start by thanking Einstein for his theory of general relativity, or the ultimate explanation of gravity as a result of time, space, and their shaping by mass. General relativity tells us that matter warps the fabric of space and time; taken to its logical extreme, general relativity predicts the possibility of places in the universe that have such highly dense matter that gravity (mostly) seals the edges and maintains a closed, compact system.

And speaking of highly dense matter . . . it's time to bring black holes into the conversation. There are two main categories of black holes. Stellar-mass black holes are formed when a massive star explodes into a supernova at the end of its life, and supermassive black holes are what lies at the center of most galaxies. And how do they work?

What goes in never comes out: How black holes work

Start with the name: Black holes. Are they actually holes? Sorry, but no; they're quite the opposite. A black hole is a region in space that is extremely dense; it's so dense, in fact, that its internal gravity prevents anything from escaping it.

What, are we going to tell you next that they're not black either? Nope, you're good on this one; black holes have such an intense internal pull that nothing — including light — breaches its edges. In this particular case, no light equals black. No formulas needed!

On second thought, let's introduce just one formula here. A good way to think about black holes is to consider escape velocity (see Chapter 2 for the details). If you throw a ball on Earth, it goes up and then comes back down due to the force of gravity. Even if you throw a ball really hard, it's going to come down eventually because the force of Earth's gravity tugs the ball back down. To get off the surface of the Earth, your ball would need something along the lines of a rocket to carry it out of the Earth's atmosphere, but we're pretty sure that league sports don't provide those.

The formula for escape velocity v_e from a large object like a planet is

$$v_e = \sqrt{\frac{2GM}{r}}$$

where G is Newton's universal constant of gravity, M is the mass of the planet, and r is the radius of the planet. Escape velocity depends on the planet's mass, and is inversely related to the planet's radius — that makes sense because gravity decreases with distance. For the Earth, if you plug in these numbers, you get v_e = about 25,000 miles per hour (11.2 km/second). A more massive (more dense) Earth would require a higher speed.

And how is this all relevant to black holes? Remember that there's a maximum speed limit in our universe — the speed of light, or 6.7×10^9 miles per hour (3×10^6 km/second). The bigger the object, the more gravity can compact it down into a smaller mass (and the denser it becomes) if fusion is turned off or never started. At some point, the object's escape velocity exceeds the speed of light, and it's at that point where nothing, not even light, can escape from it. Boom! You have a black hole.

TECHNICAL STUFF

As is the case with many things in astrophysics, the solution from Newtonian mechanics using escape velocity isn't really the whole story. Black holes arise from a solution to Einstein's theory of general relativity that updated Newton's theory of gravity to unite forces in space and time. The result? A warping or curvature of space-time. The theory is expressed in a series of field equations. German astrophysicist Karl Schwarzschild found a solution to Einstein's field equations for a special value of radius which resulted in a singularity (resulting, in this case, in dividing by zero, a physical impossibility). It turned out that this special value, the Schwarzschild radius, results when the escape velocity is set to the speed of light. (Weird thing about the universe: You can have a 5 solar mass star that is huge and supported by light from fusion, or a 5 solar mass black hole with a 9 mile (15 kilometer) Schwarzschild radius).

The event horizon: The line light cannot cross

You've heard the expression that fences make good neighbors? Black holes don't exactly come with white picket fences, but they do have boundaries. Black holes are an extremely dense aggregation of matter. Near a black hole, gravity becomes so strong that there's a point past which nothing can escape. That point, called the *event horizon*, is the edge beyond which we can't see — it's proportional to the mass of the black hole, and is also equal to the Schwarzschild radius.

It's a myth that black holes are cosmic vacuum cleaners; they don't actually go searching for innocent stars and planets to suck into their gaping mouths. Instead, their behavior is like any other massive object. Stars can orbit black holes, for example. If the stars get too close, however, some of their material can get sucked into the black hole and cross the event horizon, and if the orbit of the star is disrupted, eventually the whole star could be consumed and added to the black hole's mass. When anything (matter or electromagnetic radiation) is inside that event horizon, it can't escape except under specific conditions. (We don't even know how big the black hole is inside the event horizon!)

The critical density of a black hole is key. One way that black holes are thought to form is by the collapse of massive stars after they use up their fuel.

HAWKING RADIATION

Scientists used to think nothing could escape from a black hole, but in 1974 British astrophysicist Hawking's Stephan theorized that there could actually be a kind of runaway radiation. This radiation, called Hawking radiation, derives from a combination of quantum field theory and general relativity's curved space-time, and involves matter and antimatter particles. The basic result is that a small amount of radiation can leak out of a black hole over time. Over the age of the universe, it's possible for black holes to evaporate.

REMEMBER

Remember that although Sun-sized stars will end up as white dwarfs, larger stars will explode in a supernova and then collapse down into a neutron star. The largest stars (ones that start at about 20 times the mass of our Sun) will generally continue collapsing down to form a black hole. The minimum mass of a central core remnant (after the supernova explosion) to be dense enough to collapse into a black hole is about three times the mass of our Sun.

Seeing the invisible, or the art of detecting black holes

Black holes were predicted theoretically long before they could be observed, again thanks to Einstein's theory of general relativity. Telescopes are built to detect the wavelengths of the electromagnetic spectrum and because black holes don't give off light (and we can't detect Hawking radiation), they're very hard to detect directly. For that reason, black holes have traditionally been detected via indirect observation and inference. Scientists know a black hole leaves a footprint, for example, when it accretes (pulls inward) matter that gets too close.

Some of the most successful ways of detecting black holes include the following:

>> **Observing matter that responds to, or orbits around, a black hole's massive gravitational pull**

Even if you can't see the black hole itself, you can see objects that are moving oddly around it. Scientists used this technique to discover the supermassive black hole at the center of the Milky Way Galaxy; they tracked the motion of the stars of our galaxy that were orbiting around a patch of seemingly empty space.

>> **Watching the ripples in the space-time continuum that initiate when two black holes or a black hole and neutron star collide**

These ripples are called gravitational waves — more on those in a minute!

>> **Looking for material emitting significant heat and energy as it's pulled into a black hole**

Material that is about to be pulled into a black hole is accelerated to huge velocities. This acceleration tends to heat up the material and allow it to give off detectable radiation.

Recently, however, there have been enormous breakthroughs in the field of black hole detection. The Event Horizon Telescope (EHT) provided our first images of the surroundings of black holes, including one in the middle of our Milky Way Galaxy. Figure 8-1 shows an image of the supermassive black hole in the center of galaxy M87, outlined by a ring of bright gas and bent light that is being accelerated around the black hole. This image was created by combining observations from a network of eight radio telescopes located in different countries around the world; the signals were then combined in such a way as to make the equivalent of a telescope as big as the Earth! This is the interferometry we discuss in Chapter 4.

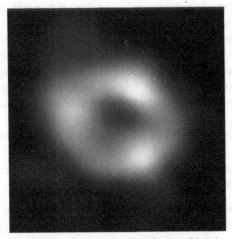

FIGURE 8-1: Supermassive black hole in the center of Galaxy M87.

Courtesy of Event Horizon Telescope collaboration

TAKING A WALK INTO A BLACK HOLE

Take a walk on the wild side, and pretend that you're strolling casually through the universe when you happen upon a black hole. You wouldn't be able to walk (or float) past it because its very strong gravitational force would pull you in, limb by limb. You'd feel the strongest pulling of your life, starting with whatever body part was nearest to the black hole, and you'd be stretched thinner and thinner. Eventually, you would be tugged over the event horizon, never to return. If that wasn't weird enough, your crewmate who was observing your demise (from a safe distance, of course), would think that time was slowing down as you were being stretched more and more slowly; black holes are so massive that their gravity can warp space-time. Conversely, to you, everything would appear to be speeding up outside the black hole. After you completely fell into the black hole, you (and everything else around you) would be compressed down to a tiny speck called a singularity. Now, does your mortgage payment really seem like such a big deal?

Surf's Up! Gravitational Waves

No trip to the ocean is complete without catching some waves . . . or, perhaps, watching someone else catch them. Or, better, waving to a friend who's out riding the waves! Waves are defined as any disturbance that transfers energy between places that can interact with other waves (and there are a few other quantum parts we don't need to worry about). For example, waves in water carry energy the water got from wind, thrown rocks, or splashing kids that exerted force on the water. Read about waves in more detail in Chapter 2.

A special kind of wave called a gravitational wave can be detected when given off by black hole collisions and similar cosmic events. These were predicted by Einstein as part of his theories of gravitation, and they're central to understanding how you can detect the invisible.

A ripple in space-time

By definition, a gravitational wave is a ripple in the space-time continuum. Tricky to understand, for sure, but let's break it down. "Space-time" is a theoretical model that combines our three linear dimensions (x, y, and z on a plot) with a fourth dimension of time. Einstein wrote about space and time as a fabric, woven together into part of the same system, and the theory of general relativity describes how gravity is a result of this continuum.

Where do gravitational waves come from? At the most basic level, a gravitational wave is created when an object is accelerated. Although some stars fade quietly at the end of their lives, others end in supernova-caliber explosions; much like throwing a stone into a pond, the effect of these explosions is thought to produce gravitational waves. Gravitational waves can also be created by a massive rotating object like a neutron star, or by two massive objects like neutron stars or black holes orbiting each other and eventually colliding and merging (the latter we already see from Earth). Figure 8-2 shows an artist's conception of gravitation waves created by the merger of two neutron stars.

Could a gravitational wave ever reach Earth? They ripple through you constantly, but you're not likely to notice it. Because the events that create gravitational waves happen light-years away from the Earth, the effect would be extremely small by the time it got anywhere near our home planet. To detect gravitational waves, you need a specially designed detector.

FIGURE 8-2:
Artist's
conception of
gravitational
waves from
neutron star
merger.

Courtesy of NASA/Goddard Space Flight Center

How to be a wave detective

Gravitational ripples result from a disruption to the space-time continuum. Indirect evidence of their existence was first found in the 1970s when astrophysicists located a binary system with two neutron stars, including one that was rotating as a pulsar. The orbit of the pulsar was found to be shrinking over time, and calculations showed that its change in orbit was consistent with the energy being emitted in the form of gravity waves.

Have scientists ever actually detected a gravitational wave, you ask? Yes, thanks to LIGO (Laser Interferometer Gravitational-Wave Observatory), a pair of huge U.S.-based interferometers in Washington and Louisiana and their sister Virgo detector in Italy. In 2015 LIGO-Virgo were able to detect gravitational waves, and determine that they'd originated from a black hole collision over a billion years ago. The discovery helped confirm Einstein's theory of general relativity which, a hundred years earlier, had predicted that if two black holes orbited each other, they would lose energy in the form of gravity waves until they eventually spiraled together and crashed into each other. As the two black holes collide and merge into one bigger black hole, a fraction of their mass is converted into energy (following $E=mc^2$) and given off as a huge burst of gravitational waves.

Neutron Stars, or Total Core Collapse

You now know all about the densest type of cosmic objects out there — black holes. Neutron stars come in at a close second on the stellar density scale. As we describe in Chapter 5, all main sequence stars spend 80 to 90 percent of their lives undergoing nuclear fusion at their cores, transitioning hydrogen into helium. If huge stars form black holes, and Sun-sized stars form white dwarfs, what about those medium-sized stars, that start out between about 8 to 20 times the mass of our Sun? They end up in the middle category of stellar end states, neutron stars.

Science of the collapse of stars

The path to a neutron star starts with a supernova explosion after the star has gone through multiple levels of nucleosynthesis, burning hydrogen and carbon, all the way up to iron (the end state of nucleosynthesis in a star). Gravitational pressure causes the core to collapse when it's no longer supported by the outward pressure of light from ongoing fusion. The supernova explosion forces outward much of the mass of the star, but what's left — the former core — has so much mass that gravity crushes the electrons and protons together to form a cloud of uncharged neutrons. Similar to the electron degeneracy pressure that supports white dwarf stars, neutron degeneracy pressure supports this 1.4-to-2.9 solar mass remnant from further collapse. Enter, stage left: neutron star!

If the star is bigger than about 20 times the mass of the Sun, the gravitational forces are sufficient to overcome this neutron degeneracy pressure and continue collapsing down, becoming denser and denser until all the mass is hidden inside a Schwarzschild radius. Neutron stars are the densest that any "normal" matter can be — and that's not the only reason to pay attention to them!

Neutron stars also have enormous magnetic fields, far greater than Earth's. The cores of neutron stars support powerful electrical currents, producing a magnetic field similar to an electromagnet — but much, much stronger! An extreme kind of neutron star called a magnetar has some of the strongest magnetic fields in the universe. The magnetic field surrounding a magnetar is so strong that it could tear apart your body at a distance of 600 miles (1000 km) by pulling the electrons off your atoms, turning your body into a cloud of disintegrated particles — not a fun way to go! Magnetars can also give off huge flares that can even affect the magnetic field of the Earth from vast distances away.

Pulsing radiation from pulsars

Pulsars are a special kind of neutron star. A pulsar has a strong magnetic field that isn't aligned with its rotational axis. As it rapidly spins, particles spray in jets from its two magnetic poles and these particles in turn emit radio radiation. Pulsars also rotate, sometimes extremely fast — on the order of hundreds of times per second, faster than the blades on your kitchen blender. As the jets get spun past us twice each rotation, we can detect pulses every few milliseconds!

As seen from Earth, a pulsar seems to "pulse" (hence the name), giving off frequent, short bursts of radiation as the beam of light passes over the Earth and then away, as shown in Figure 8-3. The effect here is similar to a lighthouse, where the rotating beam of light appears to wink on and off as seen from a distant ship. This pulsing is an illusion; the bright beam of radiation from a pulsar is always on, but we Earthlings see it in bursts as it sweeps past us.

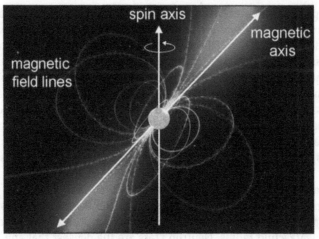

FIGURE 8-3: Pulsar diagram.

Courtesy of NASA/Goddard Space Flight Center / Public Domain

Timing of pulsars allows the calculation of many other useful astrophysical parameters, such as the masses of neutron stars in binary systems, the detection of planetary companions (see Chapter 7 for the full story), and (when data from a lot of pulsars is combined) maybe even gravitational waves from supermassive black hole mergers.

Quasars, Bursters, and Blazars

If you think your 5000-lumen flashlight is as bright as it gets, think again. Although many celestial objects are too dark to be observed easily, like black holes, others are so bright that they can be seen with the unaided eye. Blazars and quasars are both types of active galactic nuclei — disks of material with a feeding supermassive black hole at the core — and these are also some of the most luminous objects in the cosmos. Gamma ray bursts are one of the brightest sources of gamma rays around. Keep reading if you're curious as to how many quasars it takes to change a light bulb.

REMEMBER

And, if you're really curious, a 60-watt incandescent bulb gives off about 800 lumens of light. A typical quasar is trillions of times as bright as the Sun, which in turn is about 10^{25} (10,000,000,000,000,000,000,000,000) times more luminous than your lightbulb!

The XYZ of AGN: Quasars

To understand quasars, you have to start with an acronym: AGN, or active galactic nucleus. What on earth is that, you ask? Certainly nothing on earth! AGNs exist at the center of a galaxy surrounding a supermassive black hole. This disk emits jets of particles that astronomers can detect from the light that the particles emit. Quasars are the most luminous kind of AGN, and it's thought that most galaxies have at least one supermassive black hole at the center and will go through an AGN phase at some point.

TIP

The name *quasar* comes from *quasi-stellar radio source*. They were given this name because they were first discovered as objects that gave off radio waves, but looked like stars in visible light. Astrophysicists now know that they are not stars at all! Although most black holes are, well, black, those at the center of galaxies have plenty of light-making activity all around them. As black holes do, their extreme gravity wants to inhale all surrounding material. As gas and dust is sucked toward the black hole and it swirls around, it can get dense enough to heat up to extreme temperatures — hot enough to even trigger some nuclear reactions! This disk shines in electromagnetic radiation at a whole range of wavelengths from radio waves to visible light all the way up to x-rays. The result? A quasar!

Quasars are super luminous — in fact they are some of the most luminous objects in the universe, and one quasar gives off more light than the entire Milky Way Galaxy. In fact, that disk shines brighter than the entire galaxy the quasar is inside! Most quasars are located very far away — hundreds of millions of light-years — but you can still see them because they are just that bright. The fact that there aren't any quasars closer to us doesn't mean that there's something about the Milky Way that scares them away. Remember that because light travels at a finite speed, when you are looking at objects far away from us you are also looking back in time. The fact that quasars are all hundreds of millions of light-years away means that they existed and gave off their light hundreds of millions of years ago. They likely existed early in the history of the universe, but have now burned out or been blasted apart.

Blazars keep black holes in business

Some AGN (including some quasars!) point their jets right at us, and this makes them look very different from all their cousins. AGN are typically in spiral galaxies, and when we look at them from the side, we have to look through the disk of their galaxy to see them. When we look at blazars, our view isn't obscured. We also see the jet's light compressed into a smaller region on the sky than we would from the side. These factors mean blazars appear more luminous than other AGN. This perspective also makes it seem like the jets are moving apart faster than the speed of light! This is just a trick of geometry, however. Light from the far jet has to travel farther to reach us and it is emitted by objects moving away. This is tricky, but nothing — not even blazars — can break the speed of light. Figure 8-4 shows an artist's conception of a blazar.

FIGURE 8-4:
Artist's conception of a gamma-ray blazar as detected by the Fermi Gamma-ray Space Telescope.

Courtesy of NASA / M. Weiss/CfA

Where do those jets come from? As the supermassive black hole consumes gas, dust, and unsuspecting stars and planets, not all of that material falls into the black hole at the same time. Charged articles orbiting in the disk create a strong magnetic field. Particles like photons and neutrinos get wound up along the spin axis of this field and accelerated to speeds that are a decent fraction of the speed of light. This action creates the jet that is the signature feature of a blazar, one you can detect from Earth, but only if they are aligned properly for our viewing. The alignment produces a phenomenon called *relativistic Doppler boosting*, which increases their apparent brightness as seen from Earth even more.

The super-high energy neutrinos ejected from blazars have been detected by observatories like IceCube. The IceCube Neutrino Observatory is located near the South Pole and uses a volume of about a cubic kilometer of pure ice from the Antarctic Ice Sheet as a perfect neutrino detector. In addition, some blazars seem to flare up and have brief periods of time when they give off even more radiation. Blazars are thought to be an important source of cosmic rays — the charged subatomic particles that impact Earth from all different directions. Blazars can also give off gamma rays, and have been observed with the Fermi Gamma-ray Space Telescope.

Explosions from afar: Burst it out

Supernovas and blazars aren't the only evidence of violent outbursts in the universe. Gamma-ray bursts (GRBs) are rapid, very intense bursts of radiation. These are some of the most luminous and energetic events in space.

Gamma rays are a very high-energy type of radiation, and GRBs were first discovered by military satellites that were searching for evidence of secret nuclear testing during the Cold War. These brief bursts of high-intensity radiation can last from a few milliseconds to a few minutes. Longer-duration bursts are thought to be given off by special kinds of supernovae, whereas shorter-duration bursts may be created when two neutron stars merge into a black hole, or a neutron star merges with a black hole to create a larger black hole. Astrophysicists are working to simulate these events using powerful supercomputers, and additional kinds of gamma-ray bursts may yet be discovered!

3
Galaxies: Teamwork Makes the Dream Work

Learn all about galaxies — what they are, how they were found, and how they are classified, and uncover their astrophysical significance.

Explore theories of galaxy formation and galactic structure and discover why so many galaxies have a black hole at the center.

Find out more about galaxy clusters, the biggest structures in the universe; see how their physics works, and watch what happens when they merge.

Take a ride through the wormhole of galactic astrophysics and learn about dark matter halos, galactic archaeology, high energy astrophysics, and gravitational lenses.

Chapter 9

From Fuzzy Blobs to Majestic Spirals: The Milky Way and Other Galaxies

When broken down into their simplest elements, the main requirements for most celestial objects are your friends from Chapter 5: dust and gas, with a healthy serving of gravity. Stars are flaming balls of gas and dust; comets are balls of gas and dust with a tail. Gravity is the force that draws the matter in these objects inward and allows them to exist as single (or sometimes more) unit.

Now take a step back and look up. There are billions of stars visible from Earth, all united by the same gravitational system. The name for this enormous collection of entities is a galaxy, and it includes stars, more dust and gas, our solar system, and even a supermassive black hole at the center. Most of what you see in the night sky belongs to our local galaxy, the Milky Way, though ours is not the only galaxy out there. In this chapter, you learn about our galaxy and, in the process, see how other galaxies exist in conjunction with ours!

Where in the World Are We?

The Milky Way, a delicious blend of chocolate and caramel — sorry, wrong book. Although, depending on your perspective, there's nothing more delicious than thoroughly enjoying the galaxy around us. Earth's home galaxy, the Milky Way, has been visible to humans as a glowing, milky-white spread in the sky since the beginning of our collective existence. Ancient Greek astronomers wrote of a river of milk in the sky ("Gala" is the Greek word for milk), and it's likely that the origins of the name began there. Ancient Roman astronomical records then described the Milky Way as a "road of milk," the "Via Lactea," or "Via Galactica," and variants of the name continue to be used today.

You can see the disk of the Milky Way for yourself if you go out at the right time on a clear night with dark skies, far from any sources of artificial light. Get comfortable, look up, and allow your eyes to adapt to the dark; you'll soon see a bright band of stars and dust arcing across the night sky (see Figure 9-1). This arc is the disk of our own galaxy seen edge-on. As you look towards the center of the galaxy, you see more and more glow; this light is from so many stars that blur together and give the Milky Way its bright appearance.

FIGURE 9-1:
The Milky
Way Galaxy.

Courtesy of NASA

The Milky Way Galaxy isn't the oldest in the universe or even the largest. Scientists estimate its age at about 13 billion years, and it has a diameter of about 100,000 light-years.

REMEMBER

A light-year is an astronomical unit of how far light travels in one year. It's equivalent to about 6 trillion miles, or 9.5×10^{12} km. What makes the Milky Way so fascinating to Earthlings is that it's our home galaxy.

Galaxy quest

Galaxies abound in the universe. There are an estimated 2 trillion galaxies in the visible universe. Scientists have grouped galaxies largely by proximity to our galaxy, which lies in what's called the Local Group of galaxies (more on that in Chapter 11). The Local Group represents an area covering about 5 million light-years from our Milky Way's center, close enough to be neighborly by astronomical standards! The Local Group was named by American astronomer Edwin Hubble in 1936 — more on him in the later section "Hubble's puzzle."

In the early 20th century, astronomers such as Gérard de Vaucouleurs, a Frenchman working in Texas, attached more nomenclature to galaxy systems. It was soon seen that as stars could be grouped into galaxies, galaxies could also be grouped into clusters. Large clusters of galaxies are known as *superclusters*, and the Milky Way is part of a supercluster near the Virgo constellation (hence the name, Virgo Supercluster). Just like a galaxy is a gravitationally-bound set of stars, a galaxy cluster is a gravitationally-bound set of galaxies with its own structure, interactions, and evolution. More on galaxy clusters will be heading your way in Chapter 11.

Traveling the Milky Way

If you were to hop in a car (or plane . . . better still, a spacecraft) and drive down the Milky Way, you'd rapidly find yourself out of road. But what is that road shaped like?

It's hard to see something accurately from the inside out. Because we're inside the Milky Way, what scientists know about our galaxy's structure comes from any observations they can make, but also from inferences drawn from looking at other galaxies. Combining these sources of information tells scientists that our galaxy has a swirled, spiral shape with multiple arms, and that there's a giant black hole at the center. Our solar system is located out in one of the spiral arms, and the Milky Way is a barred spiral galaxy — so named because it has a central bar shape with radiating arms spiraling out of the center. Figure 9-2 shows an artist's conception of the Milky Way Galaxy.

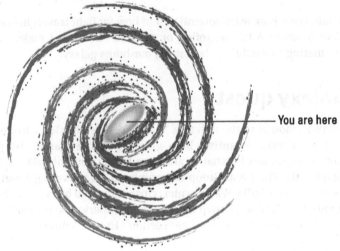

You are here

Courtesy of NASA/JPL-Caltech/R. Hurt (SSC/Caltech) / Public Domain

FIGURE 9-2:
Artist's conception of the Milky Way Galaxy.

Simply put, a galaxy is a collection of stars, planets, and other astronomical objects, such as nebulae, gas clouds, and more, all held together by the force of gravity. Galaxies vary widely in their number of stars — a small dwarf spheroidal galaxy might only have a few thousand stars, whereas a huge galaxy might have trillions of stars. Similarly, a small galaxy could be just a few hundred light-years across, while light could take a million years to cross an enormous galaxy.

Most galaxies formed early in the history of our universe, making them 10 to 13 billion years old (and those galaxies have been changing and evolving ever since). Because new stars are still forming (see Chapter 5 for more of the story), new galaxies can also form; astronomers think that some of the youngest galaxies could be only about 500 million years old. Head over to Chapter 14 for the story on galaxy formation.

Scientists think that the Milky Way is a pretty typical galaxy. It consists of about 100 billion stars, and is about 100,000 light-years across. See those spiral arms in Figure 9-3? The two largest arms are called the Scutum-Centaurus arm and the Perseus arm, and the two minor arms located between them are called the Norma arm and the Sagittarius arm. The two major arms are thought to be made of both young and old stars, whereas the two minor arms contain more gas clouds and star-forming regions.

Our Sun, and our whole solar system, is on a small jetty called the Orion Spur, located between the Sagittarius and Perseus arms. We are about halfway out from the center of the Milky Way. Of course, because everything in space is always in motion, the entire Milky Way Galaxy is also rotating; it takes the Sun about 240 million years to orbit around the center of the Milky Way once.

And what's in the middle of our galaxy? You guessed it, our very own black hole! (See more on black holes in Chapter 8.) Most galaxies are thought to have black holes at the center. In the case of the Milky Way, our black hole is called Sagittarius A* (abbreviated Sgr A*). It's a supermassive black hole, millions of times the mass of our Sun, and is located about 28,000 light-years from Earth. Astronomers recently used data from the IXPE (Imaging X-ray Polarimetry Explorer) telescope, combined with images from the Chandra X-Ray Observatory, to detect an x-ray signal bouncing off molecular clouds near the center of our galaxy. This signal was traced back to a high-energy flare from Sgr A*, which likely occurred about 200 years ago.

Unraveling the Mystery

Galaxy identification is a relatively new field; the first identification of galaxies was undertaken by French astronomer Charles Messier in the late 18th century. Initially noted as "not comets, but other fuzzy things in space," galaxies were initially classified as nebulae and other objects before Edwin Hubble arrived on the scene in the early 20th century. Applying physics and math to observation, galaxies were finally recognized and distinguished for their own distinct presence in the sky.

Galaxies and other fuzzy objects get messier

We scientists love to write down our brilliant observations and track them for posterity. Fortunately, so did French astronomer Charles Messier. Starting around 1760, he performed observations with telescopes and the unaided eye, searching for comets (he managed to discover 13 of them!). In looking for comets, though, he kept finding other fuzzy objects that didn't move from their spots in the sky from night to night (that is, not comets). He combined other astronomers' lists of nebulae (he grouped in galaxies) and star clusters and tracked his own discoveries of these annoying objects so that he didn't mistake them for comets, eventually cataloging 102 of them (8 more were added from his notes after he died). The limits of both knowledge and telescopes at the time meant galaxies like the nearby Andromeda galaxy — M31 — were catalogued as nebulae. His catalogue numbered each object and listed it with its position and a description.

Hubble's puzzle

The story of galaxies would start with Edwin Hubble, an early 20th-century American astronomer who made innumerable contributions to astronomy. One of those was expanding our perception of cosmic distance. In the 1920s, the view that astronomers had of the universe was that the Milky Way was basically all there was — the universe was thought to end at its boundaries. There were observations of weird fuzzy or spiral-shaped nebulae, but those were just thought to be clouds of gas and dust in a universe that was only our disk of stars.

But just how far away are objects in the sky? That's harder to measure than you might think! Without a sense of the true size of an object, it can be hard to tell if you're seeing, say, a small tree that's nearby, or a huge tree that is very far away. Because astronomers had no idea how far away these nebulae really were, it seemed much more reasonable to assume they were relatively small-sized gas clouds. Hubble found a way to test this hypothesis — and the results were galaxy-shaking.

Standard candles and redshifts

One way that distances have been measured in our galaxy is by using the "standard candle." This is an astronomical idea that describes any celestial object where its luminosity is a known quantity.

TIP

The main idea is that if you know how luminous something really is, its intrinsic brightness, and then you measure how bright it looks in the sky, its apparent brightness, you can compare the two and use the difference to figure out just how far away it is. A common standard candle that's used in our own galaxy and nearby galaxies is Cepheid variable stars (more on those in Chapter 5). They are bright, relatively easy to see, and, most importantly, they have a rate of pulsation that is directly linked to their intrinsic brightness. If you can identify one of these stars, and measure its apparent brightness and its pulsation rate, you can use that to calculate what its luminosity should be, and then how far away it is.

Hubble did just this when he pointed what was then the largest telescope in the world, the 100-inch telescope at Mount Wilson, at what was then called the Andromeda Nebula (M31). For the first time he resolved individual stars in M31, and then discovered that one of these stars was actually a Cepheid variable, which he called V1. Hubble's calculations on the distance to V1 showed that it was vastly farther away than any other object in our own galaxy, and in fact that Andromeda was a separate galaxy in its own right — an island universe — rather than part of the Milky Way! Figure 9-3 shows observations of V1 taken by Hubble's namesake, the Hubble Space Telescope, tracking its variations.

An example of a mid-level solar flare, a powerful burst of radiation from the surface of the Sun, as captured by NASA's Solar Dynamics Observatory. Learn all about the Sun and how space weather affects us on Earth in Chapter 4.

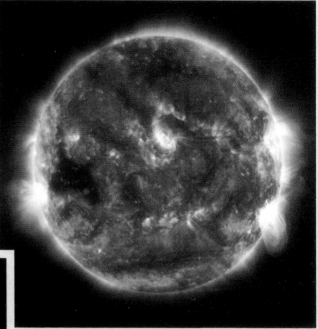

The eight planets of our Solar System with the Moon. From the top in this montage, the four planets of the inner Solar System are: Mercury, Venus, Earth (with Moon) and Mars. They are roughly to scale with each other, and the four planets of the outer Solar System (Jupiter, Saturn, Uranus, and Neptune) are roughly to scale with each other. The inner and outer Solar Systems are not at the same scale. See Chapter 3 for more on our Solar System.

Jupiter's moon Europa, in a mosaic of images from the Galileo spacecraft. Europa is about the size of Earth's Moon, but underneath its icy surface lies an ocean of liquid water larger than Earth's. Learn more about that search in Chapter 7.

In an annular solar eclipse (left) the Moon covers nearly all of the Sun, leaving a bright ring in its wake. In contrast, a total solar eclipse (right) occurs when the Sun is completely covered by the Moon which reveals the Sun's surrounding corona. During the brief minutes of a total eclipse, it's safe to look at the Sun without eye protection — but not the case during an annular eclipse! Chapter 4 has more on all kinds of eclipses.

An artist's conception of exoplanet system LTT 1445A. Earth-sized exoplanet LTT 1445Ac is visible as the dark spot crossing the disk of a red dwarf star, one that's located in a triple system with two other red dwarfs at the upper right. Another exoplanet, LTT 1445Ab, is shown in the foreground. Learn about exoplanets in Chapter 7.

Globular cluster Messier 15, in the constellation Pegasus, is seen in this image from the Hubble Space Telescope. Cool yellow stars and hotter blue stars are visible in this 12 billion-year-old cluster. For more on star clusters and their friends, see Chapter 6.

The star Wolf-Rayet 124, as seen by the James Webb Space Telescope (JWST) in near-Infrared and mid-Infrared wavelengths. This extremely bright star is in the process of shedding its outer layers (visible as the gas and dust surrounding the central star) before it explodes in a supernova. Chapter 5 has all the details on the life cycle of a star.

The Crab Nebula, a supernova remnant, as seen in near infrared and mid infrared by JWST. The supernova explosion containing a bright pulsar at the center was first recorded by Chinese astronomers in 1054. Chapter 6 is your source for more on nebulae.

A star forming region in Rho Ophiuchi, as seen by JWST. Dark areas are dust cocoons around still-forming protostars, and newly-revealed stars shoot off as red bipolar jets of molecular hydrogen. See Chapter 5 for details on gas, dust, and star formation.

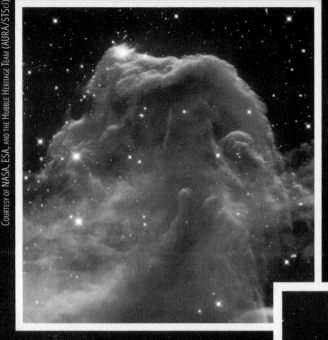

Courtesy of NASA, ESA, and the Hubble Heritage Team (AURA/STScI)

An infrared view of the Horsehead Nebula as seen by the Hubble Space Telescope. The infrared view penetrates the outer, dark dusty layers of the nebula to reveal its interior and the new stars forming within. For more, see Chapter 6.

A protostar embedded in a dark cloud of dust and gas is visible in the infrared in this image from JWST. The central star, in the dark cloud L1527 in the Taurus star formation region, has shocked the nebula above and below it; the result is thinner regions that glow in blue and orange. Get a head start on star formation with Chapter 5.

Courtesy of NASA, ESA, CSA, and STScI. Image processing: J. DePasquale, A. Pagan, and A. Koekemoer (STScI)

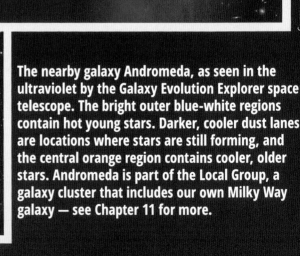

Courtesy of NASA/JPL-Caltech

The nearby galaxy Andromeda, as seen in the ultraviolet by the Galaxy Evolution Explorer space telescope. The bright outer blue-white regions contain hot young stars. Darker, cooler dust lanes are locations where stars are still forming, and the central orange region contains cooler, older stars. Andromeda is part of the Local Group, a galaxy cluster that includes our own Milky Way galaxy — see Chapter 11 for more.

Barred spiral galaxy NGC 1300, as seen by the Hubble Space Telescope. Visible in the spiral arms are blue and red supergiant stars, as well as star clusters and star formation regions. The central bar shows the galaxy's spiral structure traced out by darker dust lanes. Chapter 9 has the details on galaxy classification.

Courtesy of NASA, ESA, and The Hubble Heritage Team (STScI/AURA); P. Knezek (WIYN)

Courtesy of NASA, ESA, CSA, STScI

A view of Stephan's quintet, a grouping of five galaxies (four of which are gravitationally bound into a cluster, while the left-most is in the foreground), as seen by JWST in the mid and near-infrared. Red and gold regions show the shock waves generated as galaxy NGC 7318B, at the center of the image, smashes through the rest of the galaxy cluster. See Chapter 11 for more on the physics of galaxy clusters.

Courtesy of NASA, ESA, S. Baum and C. O'Dea (RIT), R. Perley and W. Cotton (NRAO/AUI/NSF), and the Hubble Heritage Team (STScI/AURA)

Jets of material ejected due to a supermassive black hole at the center of the galaxy Hercules A are seen in this multi-wavelength combination image using data from the Hubble Space Telescope and the Very Large Array radio telescope in New Mexico. The red jets in the image are invisible to HST but are very powerful radio sources. Chapter 10 has more on active galaxies.

Galaxy group NGC 5813 glows in this composite view made from X-ray observations taken by the Chandra X-ray Observatory (shown in purple) combined with optical data. A spinning supermassive black hole at the center of the galaxy group expels gas and dust, and its interacting shock waves generate jets and voids visible in the image. Chapter 12 has the details on this and other high-energy astrophysical phenomena.

Courtesy of NASA/CXC/SAO/S. Randall et al./SDSS

Courtesy of X-ray: NASA/CXC/IIA/INAF/J.Merten et al, Lensing: NASA/STScI; NAOJ/Subaru; ESO/VLT, Optical: NASA/STScI/R.Dupke

Galaxy cluster collision Abell 2477, also known as Pandora's Cluster, as seen in this composite image. Data from the Chandra X-ray Observatory reveals hot gas (red in the image), while blue regions indicate the mass concentration (primarily dark matter) mapped using data from the Hubble Space Telescope as well as ground-based observatories. See Chapters 11 and 15 for more on dark matter.

Astronauts Richard Linnehan and John Grunsfeld on a spacewalk to repair and upgrade the Hubble Space Telescope with a new cryogenic cooler for one of its near-infrared instruments. The Hubble Space telescope is visible in the background, temporarily attached to the Space Shuttle Columbia's cargo bay for the repairs. Chapter 18 has details on this and other important space telescopes.

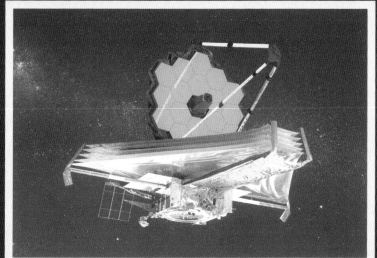

Artist's conception of the James Webb Space Telescope, with its segmented mirror (yellow) and sunshade (purple) deployed. JWST orbits at the Lagrange point L2, a stable point in the Earth's orbit around the Sun, and the sunshade is used to provide passive cooling for the telescope's near-infrared instruments. See Chapter 18 for more info.

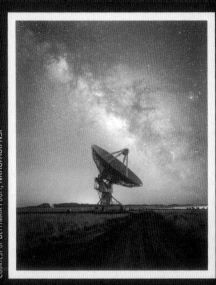

The disk of our Milky Way galaxy is visible arching over a dish from the Karl G. Jansky Very Large Array (VLA) in New Mexico. The VLA consists of 27 identical radio telescope dishes, each with a parabolic antenna that is 82 feet (25 meters) across. See Chapter 4 for more on radio telescopes and other kinds of ground-based telescopes.

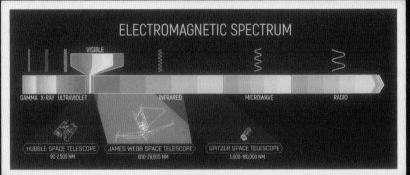

ELECTROMAGNETIC SPECTRUM

VISIBLE

GAMMA X-RAY ULTRAVIOLET   INFRARED   MICROWAVE   RADIO

HUBBLE SPACE TELESCOPE   JAMES WEBB SPACE TELESCOPE   SPITZER SPACE TELESCOPE
90-2,500 NM   600-28,500 NM   3,000-160,000 NM

Electromagnetic radiation spans a range of wavelengths from high-energy gamma rays and x-rays to ultraviolet, visible light, and longer wavelengths such as infrared, microwave, and radio waves. The electromagnetic spectrum displays this radiation in order of wavelength. This diagram indicates the wavelength ranges covered by the Hubble Space Telescope, the James Webb Space Telescope, and the Spitzer Space Telescope. Chapter 1 has more details on the electromagnetic spectrum.

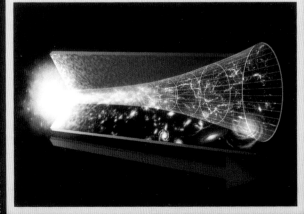

An artist's conception of the evolution of the universe. The diagram starts with the Big Bang on the left and shows the formation of matter, stars, and galaxies in the modern era. The red arrow at the bottom indicates the direction of time. Learn more about our cosmic origins in Chapters 13 and 14, and the ultimate fate of the universe in Chapter 16.

FIGURE 9-3:
Cepheid variable star V1 in the Andromeda Galaxy M31.

Courtesy of NASA, ESA, and the Hubble Heritage Team (STScI/AURA)

REMEMBER

Another important aspect of celestial objects, including galaxies, is their Doppler shift. Recall from Chapter 4 that redshift takes place when a light wavelength stretches, or shifts, toward the red portion of the light spectrum as an object moves away. The Doppler effect (with additions for air) also happens to sound. Ever listened while a train passes you at high speed? The pitch of the wailing train gets higher as it approaches you and the sound waves are squished together, and then deepens as it passes you because the sound waves get stretched out as the train moves away from you.

TECHNICAL STUFF

THE REDSHIFT EQUATION

The Doppler effect calculates the speed of an object, v, compared to the speed of light, c. Astronomers generally don't worry about v, but instead look at the ratio and call that redshift, z:

$$\frac{v}{c} = z$$

To calculate redshift, you need to compare the wavelength of some well-known emission lines in a laboratory, such as those that can be measured from hydrogen, with the observed spectrum of a star or a galaxy. There's a pattern of four fundamental lines given off by hydrogen at different wavelengths, that are always at these wavelengths

(continued)

(continued)

when observed here on Earth. If you observe a galaxy, you find this pattern of 4 lines with the same distance between the lines, but the whole pattern may be shifted to the blue or the red along the spectrum. This is a redshift or a blueshift. The redshift, which is usually referred to as z, can be defined as:

$$1 + z = \frac{\lambda_{observed}}{\lambda_{rest}}$$

Here $\lambda_{observed}$ is the measured location of one of the lines in a celestial object, and λ_{rest} is what was observed in the lab (where things should not be moving). If z is positive, you have a redshift, and the object is moving away from you, and if it's negative, you have a blueshift and the object is moving towards you. A large backyard telescope might be able to see a galaxy with a redshift $z = 0.1$. You can then convert that into velocity, to figure out how fast the object is moving towards or away from you, with the simple formula $v = c z$ where c is the speed of light. So, if $z = 0.1$, then the object's velocity is about 6.7×10^7 miles per hour (30,000 km/second). This means that the physical meaning of z is velocity as a fraction of the speed of light. One caveat: this simple relationship only works for objects at relatively low redshift — after you get close to $z = 1$, things get a lot more complicated thanks to relativity!

As if Hubble's revelation that separate galaxies existed beyond our own wasn't enough, he also discovered a weird relationship between redshift and distances. As Hubble studied the distant galaxies he helped discover, he was able to measure their spectra using telescopes, and determine their redshifts. Hubble found that most galaxies were in fact moving away from us, and, even weirder, the farther away a galaxy was (determined using his Cepheid variable trick and other methods), the higher its redshift and so the faster its velocity. Today, we know the only blueshifted galaxies are in our own local group, and orbit with us.

This fundamental relationship between distance and redshift is now called Hubble's Law, and it is such an astronomical foundation that these days, redshift is used as a measurement of distance — high redshift galaxies are much farther away from us than lower redshift galaxies. Why are all those galaxies moving away from us? No, it's not that tuna salad sandwich you ate for lunch — in fact, the galaxy motion was a serious clue that our universe is expanding. Hubble's Law lays the groundwork for the Big Bang and our study of modern cosmology — more on that in Chapter 13.

Galaxy Classification

Now that you've got a good handle on galaxy composition, detection, and measurement tools, we can talk about sorting. Without a means of categorizing galaxies into types, it would be impossible to draw useful inferences about their history, significance, and their effect on the evolution of the universe. Fortunately, galaxies share enough commonalities that they can be clearly grouped (with, of course, a few exceptions).

In a burgeoning 20th-century era of galactic discovery, there was a tangible need to classify galaxies into similar groupings for analysis and comparison. Stepping up to meet that need was none other than Edwin Hubble (yet again!). His 1926 creation of a galaxy classification system revolutionized the way in which astronomers identified galaxies by shape and size.

TIP

Hubble's scheme, often called the "Hubble tuning fork" because its shape resembles that of a musical tuning fork (see Figure 9-4), organizes galaxies as Elliptical and Spiral. See Table 9-1 for more information on each of these categories.

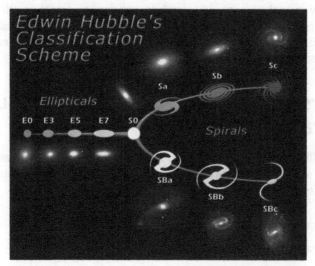

FIGURE 9-4:
The Hubble
tuning fork
diagram.

Courtesy of NASA & ESA / Public Domain

TABLE 9-1

Galaxy Types Defined by Hubble

Galaxy Type	Hubble Identification Prefix	Star Age
Spiral galaxy	Sa, Sb, Sc (Type A has the tightest-wound arms, Type C has the loosest-wound arms) SBa, SBb, SBc (same as above for barred spiral)	More younger than older stars
Elliptical galaxy	E	Older stars
Lenticular galaxy	S0 (SB0 for barred lenticular)	Older stars
Irregular galaxy	Irr	Variable mix of younger and older stars

The Spiral category branches into two arms to separate barred spirals from regular spirals, and lenticular and irregular galaxies float in the central space between these two major types. Within each type, subcategories reflect how round or sausage-shaped an elliptical galaxy is, and how loosely or tightly wound the arms of a spiral galaxy are. Hubble originally intended the diagram in Figure 9-4 to show an evolutionary sequence from one type to another, with an orderly progression from elliptical to spiral. Astronomers today know that galaxy evolution is a lot more complicated than this simple scheme, but the categories are still useful. See Chapter 10 for more about galaxy evolution.

Stars and gas with uncommon beauty: Spiral galaxies

Our Milky Way Galaxy is an example of a spiral galaxy, so named because — well — spiral galaxies are shaped like a spiral, or pinwheel. All spiral galaxies have the following:

>> A central bulge portion made up of older stars

>> A star-, gas- and dust-filled rotating central disk

>> Round halo regions surrounding the disk plane

>> Two or more arms, made up of gas and stars, that spiral out from the central region

So where do these spiral arms come from? All galaxies rotate around their center, and a spiral galaxy experiences what's called differential rotation — that means that everything in the galaxy doesn't move in lockstep, like the horses on a carousel. Stars' velocities are determined by how much mass and distance is between them and the center of the galaxy. Because distances are shorter near the galactic center, stars near the middle might have already completed one or multiple orbits in the same time it took stars near the outer portions of the galactic disk to make it halfway through one orbit.

Any areas with a little extra mass in this disk can gravitationally cause what is called a spiral density wave. As the stars, gas, and dust of a galaxy move through an area with extra mass, they bunch up and linger. In systems with a bar of stars of a neighboring galaxy, these structures can set up waves that create the appearance of bright well-defined arms. In addition, the spiral arms are gassier (no one said galaxies were polite) and filled with younger stars than the wiser and more experienced stars that make up the halo and bulge areas. This is because the density wave causes gravity to push together gas and dust to form new stars more efficiently in the spiral arms than elsewhere in the galaxy, so the arms are full of young stars and star-forming regions. Figure 9-5 shows a beautiful spiral galaxy, M100, which has bright star formation regions in its arms.

FIGURE 9-5:
Spiral galaxy M100, as imaged by the Hubble Space Telescope.

Courtesy of NASA, ESA and Judy Schmidt

A subset of the spiral galaxy is called a barred spiral galaxy. These more mature galaxies are characterized by bars, or ribbons, of gas and star material through the center. Astronomers think that galaxies start out as spirals and then evolve over time into barred spirals, and that earlier in the history of the universe, barred spiral galaxies were much less common than they are today. For more on galaxy formation and evolution, see Chapter 10.

Some barred spiral galaxies are even more complicated, such as the grand-design spiral galaxy NGC 1300, as shown in Figure 9-6. If you look closely at the center of the bar, you can see an inner disk with another "grand design spiral" structure. This structure shows the bar is funneling material toward the galaxy's core, although NGC 1300's central black hole doesn't appear to be feeding on it.

Courtesy of NASA, ESA, and The Hubble Heritage Team (STScI/AURA); Acknowledgment: P. Knezek (WIYN)

FIGURE 9-6:
Barred spiral galaxy NGC 1300, as imaged by the Hubble Space Telescope.

Never lose foci: Elliptical galaxies

Elliptical galaxies look like dinosaurs. That would be a true statement, were dinosaurs shaped like ellipses. Elliptical galaxies sometimes are almost round, but most have an ellipsoid or sausage shape. Generally speaking, elliptical galaxies:

>> Are shaped anywhere from almost-spherical to very elliptical

>> Have stars which orbit around the center more randomly than other galaxies

>> Are often found inside of galaxy clusters

>> Consist of older stars and are redder than other kinds of galaxies

>> Contain less dust and gas than spiral galaxies

Elliptical galaxies have less gas and dust than spiral galaxies, and therefore tend to mostly consist of older stars because they no longer have active star formation regions. Elliptical galaxies also don't have much structure — they are just a big football-shaped mass of stars. Most of the stars are older, redder stars that are cooler than the young blue stars found in spiral galaxies. Giant elliptical galaxies also tend to be surrounded by a halo of globular clusters (see Chapter 6), another indicator of age (and the consumption of other galaxies).

TIP

Some elliptical galaxies are formed through mergers with other galaxies, and have the structure to prove it. Galaxy NGC 2865, for example, shown in Figure 9-7, is thought to have formed when a spiral galaxy merged with a smaller elliptical galaxy. This collision started a new era of star formation in the newly merged galaxy, meaning that NGC 2865 has a higher proportion of younger stars than a usual elliptical galaxy. Figure 9-8 also shows how the galaxy is surrounded by a shell of gas — this halo was created during the galaxy merger, as well, from gas that was ejected.

FIGURE 9-7:
Elliptical galaxy
NGC 2865, as
seen by the
Hubble Space
Telescope.

Courtesy of ESA/Hubble & NASA; Acknowledgement: Judy Schmidt

Lenticular galaxies are shaped like . . . guess what?

In between elliptical and spiral galaxies lies Hubble's family of lenticular galaxies. These are characteristically lens-shaped, flat at the edges and bulging up in the middle, and they have properties of each of our previous galaxy types:

>> Lenticular galaxies have the same kind of disk and central bulging region as spiral galaxies, but they lack defined spiral arms.

>> Barred lenticular galaxies contain ribbons or bars as seen in spiral barred galaxies.

>> Similar to elliptical galaxies, lenticular galaxies contain relatively older stars.

TIP

If you see a galaxy face-on, it can be hard to tell if it's an elliptical or a lenticular because you need a bit of a side view to see the diagnostic central bulge. Galaxy NGC 6861, shown in Figure 9-8, is a classic example of a lenticular galaxy. The dark bands of dust are called *dust lanes,* and are helpful to figure out if you are seeing a galaxy from the side or the top — in this case, the dust lanes show that you're seeing NGC 6861 from an angle, and that it indeed has a central bulge and no arms — bingo, a lenticular galaxy has been found!

FIGURE 9-8:
Lenticular galaxy NGC 6861, as seen by the Hubble Space Telescope.

Courtesy of ESA/Hubble & NASA; acknowledgement: J. Barrington

Because lenticulars seem to have older stars than spiral galaxies, it's possible that they form from spiral galaxies whose arms have faded, or they could form when two spiral galaxies merge at just the right angle. NGC 6861 is part of a group of about a dozen galaxies in the constellation Telescopium. Because being in a galaxy group increases the chances of mergers, that could point towards the "merging galaxies" hypothesis. Generally, lenticular galaxies seem to have used up most of their gas and dust, and are no longer forming new stars — instead, they are just peacefully living out their days (and this could be a very long life).

Last night's leftovers on a universal scale: Irregular galaxies

Although spiral, elliptical, and lenticular galaxies are some of the most common large galaxies, let's not forget about the geometric outliers that don't fit into Hubble's nice classification scheme. Irregular galaxies account for two to four percent of total galaxies, and don't fit cleanly into either spiral or elliptical galaxy shapes; rather, they can be shaped like doughnuts, loops, or any number of variants. They also vary greatly in size, ranging from 100 to 10 billion times the mass of the Sun. It's thought that irregular galaxies could be created from collisions of existing galaxies, or perhaps represent an in-between period in a galaxy's life where it's transitioning from one classification to another.

Irregular galaxies can have both young and old stars, usually have high amounts of gas, and can contain star formation regions. Two famous irregular galaxies are the Small and Large Magellanic Clouds, which are satellite galaxies of the Milky Way.

IN THIS CHAPTER

» **Using galaxy formation to understand the cosmos**

» **Demystifying the physics of massive star systems**

» **Taking inventory on a galactic scale**

» **Diving into the dark center of a galaxy**

» **Expanding your knowledge of an expanding universe**

Chapter **10**

Quantifying the Unknown, or How Galaxies Work

Remember your best friends from Chapter 9, dust and gas? At a high level, you're already in the loop regarding their role in creating galaxies. After the universe came into being via the Big Bang (we explore this topic in more detail in Chapter 13), space was filled with gas and cosmic radiation. The rotation and collapse of the gravity-fueled clouds led to the emergence of stars that rapidly lived and died and enriched the universe with heavier atoms. These heavier atoms led to dust, the gas and dust created more stars, and these stars evolved into star clusters and eventually became the disk and halo that indicate the formation of a galaxy.

There's more to galaxies than meets the eye, in both character and quantity; after all, it's not like you can look out your back door and see the trillions of galaxies that exist in the universe! Galaxies are an amazing microcosm of astrophysics in that they encapsulate the inner workings of time and space. They are the complete package. From black holes (check!) to star formation (check!) to nature's fundamental forces (check!), galaxies run the gamut of all that astrophysics has to offer.

Galaxy Formation Helps Unravel the Cosmos

Is ours the only galaxy out there? No (Milky) way. Earth- and space-based observations have predicted the existence of billions of other galaxies, all at varying distances from our own. Understanding galaxies provides direct insight into the rate of star formation in the universe. Stars have been forming at a mostly constant rate over time since the first stars formed, about 200 million years after the Big Bang, and it took another few hundred million years for them to gather together into the first galaxies. We talk more about these first-generation stars and galaxies in Chapter 14, but for now let's probe a bit more deeply into a few of the fundamental aspects of galaxy creation.

Source matter: No EM radiation for you!

TECHNICAL STUFF

Matter is, by definition, the material that makes up the universe. The Earth is made of matter — 5.97219×10^{24} kg worth of oxygen, iron, aluminum, and other elements — as are planets, stars, and everything else out there.

Galactic source material is — surprise! — matter, but matter of a particular sort. Gas and dust, sure, but also stars, dark matter, and even black holes make up the composition of a typical galaxy. And what exactly is dark matter?

REMEMBER

Remember Hubble's discovery in Chapter 9: All galaxies seemed to be moving away from us, and the farther away they were, the faster they were moving. That idea led to the discovery of the expansion of the universe. We talk about this more in Chapter 13, but the key point here is that astronomers' ideas about how the universe formed and what's happened from there require more mass and energy than can be seen in the observable universe, and that's where dark matter comes into play.

Dark matter is thought to make up about 27 percent of the universe. The "dark" part means that this kind of matter can't be detected with any of the usual techniques, but it does exert a gravitational force on other matter. Dark matter doesn't give off any electromagnetic radiation either, so it can't be seen directly with an x-ray telescope (or any other kind of telescope!) If you think dark matter is weird, wait until you hear about dark energy; this is another concept we demystify in

Chapter 15. For now, let's focus on the "normal" matter that makes up galaxies, also known as detectable matter.

What is known about galaxy formation? The prevailing theory is that galaxies have formed and matured over time, likely starting as elliptical galaxies but perhaps colliding and merging into completely different shapes.

REMEMBER

As you look for galaxies at higher redshifts (more details are in Chapter 9), you're looking back in time at galaxies that formed long ago, when the universe was younger. These ancient galaxies look very different from today's majestic spirals and ellipticals; ancient galaxies are comparatively lumpy, for lack of a better word, with knots of bright material where stars are forming at a rapid pace.

The trinity: Gravity dominance, instability, collapse

If you think the main forces at play in your life are love and physical attraction, you're half right. Without gravity, the attractive force that pulls all objects in towards its center, you'd be floating away from the surface of the Earth (and the Earth wouldn't be pulled into orbit around the Sun, but that's another problem entirely). Gravity plays a central role in your everyday life, and the same is true for galaxies.

Of the major forces at play within a galaxy, gravitational dominance and instability top the list. We cover gravitational instability as it relates to dark matter in Chapter 11.

TIP

Tired of hearing about dust and gas in the universe? Sorry, but not sorry — the formation of stars, the universe, and everything in between all begin at the beginning; cosmically speaking, everything always comes back to primordial dust and gas. When it comes to creating galaxies, though, you can add another level of detail. Clumps of gas and dust begin to hold together (in piles of dark matter), until the density increases enough to collapse. This concept is similar to the way stars form, but on a much larger scale!

TECHNICAL STUFF

One of the main ideas about how galaxies form explains that when a cloud of gas and dust gets large enough, gravity may overcome gas pressure, causing the cloud to collapse. If there is any asymmetry (and there is always asymmetry!), the system will start to rotate. If the collapse is rapid, stars may form with chaotic orbits, resulting in an elliptical galaxy. If the collapse is slower, a disk galaxy will have a chance to form. This idea is called the theory of gravitational collapse for galaxy formation.

The gas that's caught up in this formation process gets heated due to shock waves, and then must cool according to the laws of radiative physics. The cooling rate depends on the temperature and pressure of the gas, but it happened fairly efficiently in the hot, dense early universe when many galaxies were first forming.

Final push: Gravitational torque with a side of angular momentum

Do you ever feel the Earth's rotation when you're standing outside? No, neither do we. The Earth rotates about once every 24 hours, and a person at the equator travels at what seems like an alarming pace of 1000 miles per hour (1600 kilometers per hour). However, you're rotating right along with the Earth, so you don't notice (do you?). Similarly, all galaxies rotate and our Milky Way Galaxy is no exception. It spins at about 130 miles per second but because you're inside the galaxy, you don't feel the effects of its motion.

REMEMBER

Why do galaxies experience rotation? Look to angular momentum for the answer. As you learn in Chapter 2, linear momentum is a property of objects that have mass and are moving in a particular direction. Angular momentum adds in the idea that an object's position and the direction of its motion changes over time, as compared to a fixed reference point. If you slide a block down a slope, it has linear momentum. If you attach that block to a string and swing it around your head, it also has angular momentum (just don't hit anyone!)

TECHNICAL STUFF

Additionally, scientists use gravitational torque to model exactly how fast galaxies are rotating and how infalling gas gets incorporated. The term *torque* (also called *moment*) in physics refers to the force required to rotate an object around its axis. Torque is described by this equation:

$$\tau = rF\sin\theta$$

where τ = torque, r = distance from the center of mass to where the force is applied, F = the applied force, and θ = the angle between the force and the lever arm (the perpendicular distance between the line of force and the rotational axis). Add gravity as that force into the mix and bam! You have gravitational torque, or the torque due to gravity, as shown in Figure 10-1. You can understand gravitational torque as the force of gravity bearing on a mass's center of gravity, causing it to rotate around some fixed point. For instance, if you hold a pencil horizontally by its eraser and loosen your grip so it can swing, torque will rotate it around the eraser you're holding. Scientists apply the concept of gravitational torque to help understand how gas particles can lose their spin and join in the rotation of a galactic disk.

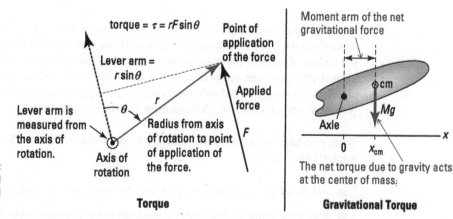

$$\text{torque} = \tau = rF\sin\theta$$

Point of application of the force

Lever arm = $r\sin\theta$

Lever arm is measured from the axis of rotation.

Radius from axis of rotation to point of application of the force.

Axis of rotation

Applied force

F

Torque

Moment arm of the net gravitational force

cm

Mg

Axle

0   x_{cm}

x

The net torque due to gravity acts at the center of mass.

Gravitational Torque

FIGURE 10-1: Torque and gravitational torque.

What does all this physics have to do with galaxy formation? As galaxies form, the collapsing core is subject to forces from nearby clumps of matter. This force, plus the original motions of the gas and dust that collapse down, results in a spinning disk of material due to conservation of angular momentum.

Mechanics of a Star System

If you think those star clusters in Chapter 6 are complicated, now think about a galaxy with billions of stars! Kind of mind-blowing, isn't it? Here in the world of astrophysics we try to blow your mind at least once a day, so consider this your daily dose.

TIP

Our Sun and our whole solar system orbit the center of our Milky Way Galaxy. You can use the orbital velocity of the Sun, about 200 kilometers per second, to find that the solar system's orbital period is about 225 million years; that's how long it takes us to take a single trip around the center of the galaxy. To put that into perspective, over the whole time that the Earth has existed, it's only orbited the center of the galaxy about 20 times.

REMEMBER

Think of Kepler's third law (covered in Chapter 2), when you have one object orbiting another. If you assume that most of the mass of the galaxy is concentrated at the center, or at least inside the orbit of the Sun, basic calculations estimate the mass of the galaxy at about 100 billion solar masses. This assumption was thought to be reasonable for quite some time, because most of the visible material in the Milky Way seems to drop off dramatically after you get out to the edges. This idea, however, is quite false.

Kepler's third law also shows that objects orbiting a massive central object at a large distance will move more slowly than those closer to the mass. In the case of our galaxy, if most of the mass is concentrated in the visible portions of the galaxy, objects that are seen to be farther out from the Sun (such as globular clusters) should be moving more slowly than the Sun. In fact, however, those globular clusters have orbital velocities that are even faster than the Sun!

What you have on your hands, folks, is a conundrum. Objects move according to Kepler's third law:

>> If they are going too slow, they fall into the center of the system.

>> If they are going too fast, they escape from the system completely.

TECHNICAL STUFF

If these outer globular clusters are moving faster than expected but are also bound to the system and not escaping, there must be significantly more gravity (and therefore more mass) in the galaxy than can be seen. Calculations show that you need about 20 times more mass — 2000 billion solar masses — in our galaxy (or at least inside the orbit of the globular clusters!). Because the material can't be sensed with any of our normal ways of studying electromagnetic radiation, that means it must be — you guessed it — dark matter. More on this in Chapter 15.

Galactic Structure

Thank you, Big Bang. That's almost all we need to say about galactic structure.

TIP

Almost, but not quite. Everything in the universe stems from the Big Bang, including galaxies (more on this in Chapter 13). The initial period of expansion in the universe was full of hot air, literally — gas, predominantly helium and hydrogen, filled the available space until stars began to form. Stars clustered and led to galaxies — or did the galaxies come first? It's a bit of a chicken-and-egg problem but, predictably, there are several different theories about the science behind galaxy formation.

The three main galaxy formation theories are

1. The classic theory of galaxy formation is **top-down:**

   - A huge spinning disk of material forms and collapses.

   - The disk eventually fragments and clumps to create individual stars.

   - Stars form in place from a large, gassy proto-galaxy.

2. The second "stars-first" theory is more **bottom-up:**

- This idea starts with individual stars that form in nebulae or star formation regions.

- Eventually enough stars form that they become gravitationally bound into a star cluster, and finally a galaxy.

3. The third and most recent theory relies on the idea of **mergers:**

- Early galaxies were small, clumpy affairs with lots of dark matter.

- Galaxies grew to their current size through repeated mergers, becoming larger and growing more complex structures over time.

- A burst of star formation is triggered every time galaxies merge.

TIP

The galaxy-merger theory is currently the most widely accepted for explaining the majority of galaxies, though it appears some galaxies did grow through combinations of the three mechanisms. Massive galaxies in the early universe seem only possible through top-down growth. Galaxy mergers of all sizes are certainly seen in the universe. Figure 10-2 shows a galaxy merger, as seen in a combination of data from the Spitzer and Hubble Space telescopes. A burst of bright, hot material at the center of this image is a location with very strong star formation that's driven by the merger.

FIGURE 10-2:
Merging galaxies, as seen in a combination of data from the Spitzer Space Telescope and the Hubble Space Telescope.

Courtesy of NASA/JPL-Caltech/STScI/H. Inami (SSC/Caltech)

REMEMBER

There's an old adage that a snake can shed its skin over and over, but it's still a snake. Galaxies travel a serpentine path here because in their initial formation, galactic structure is not a constant. Recent studies using data from the James Webb and Hubble Space telescopes were able to time-travel, in a sense, to observe galaxies farther and farther away from Earth. Objects located way out in left field — or right field, or any field, as long as they're far away — are older than nearby objects thanks to the finite speed of light (see Chapter 2). These studies showed that more stars existed in elliptical galaxies in eons past than they do today, indicating that the fundamental shape of these galaxies has changed over time.

An inventory of the parts

We take a short break here from "everything is a ball of dust and gas" and break galaxies down into their specific parts. Chapter 9 discusses the four main types of galaxies according to their shape classification: spiral, elliptical, lenticular, and irregular. The common components in galaxies include their disk, central bulge, and sometimes arms and halo. See Table 10-1 for a quick reference.

TABLE 10-1   **Parts of a Galaxy**

Type	Central Bulge	Disk of Stars	Halo	Spiral Arms	Bars	Central Black Hole
Spiral galaxy	Yes	Yes	Yes	Yes	Yes, if a barred spiral galaxy	Yes
Elliptical galaxy	Yes	No	Yes	No	No	Yes
Lenticular galaxy	Yes	Yes	Yes	No	No	Sometimes
Irregular galaxy	No	Yes	Sometimes	No	No	Sometimes

The different components of a galaxy include

» **Central bulge:** Contains the highest density of stars; some are young and hot, but mostly it consists of older stars. Density of the Milky Way's central bulge is a million times greater than that near our Sun.

» **Disk:** Contains most of the gas (hydrogen and helium) and dust. In a spiral galaxy, dark dust lanes are visible between bright spiral arms. Hot young stars are closer to the galactic plane, whereas older stars like the Sun may be more above it.

>> **Halo:** Consists of some globular clusters, has little gas or star formation, and contains primarily dark matter. The Milky Way's halo may contain a trillion times the mass of the Sun in dark matter, and extends out in a sphere at least 1 million light-years from the center. The vast majority of a galaxy's mass is in this halo.

>> **Spiral arms:** Regions of higher density of bright stars; a spiral density wave creates the spiral appearance. A *bar* (not always present) is a central linear region from which spiral arms originate.

>> **Central black hole:** Most galaxies have a supermassive black hole at their center, though small dwarf galaxies may lack one. Some flat or lenticular galaxies that lack a central bulge also may lack a supermassive black hole.

TECHNICAL STUFF

Galaxies can have different populations of stars in different locations. A typical spiral galaxy, for example, might have younger, higher-metal-content stars (Population I) in its spiral arms, but could also have older, lower-metal-content stars (Population II) in the halo. This distribution is one clue that the galaxy could have started out as an elliptical galaxy with a cloud of older, low-metal stars, and over time began spinning faster. This motion would have concentrated most of the gas and dust in the disk, where more recent stars began forming. For more on different populations of stars, see Chapter 14.

Stellar content of galaxies

There are some predictable aspects of astrophysics. Planets, spacecraft, and astronauts alike all obey Newton's Laws, for example (as long as they aren't moving at a speed close to the speed of light, because all bets are off there). Kepler's Laws of Motion that describe planetary orbits apply consistently to planets across our solar system, planets around other stars, and even to the motions of stars in a galaxy. Not everything in space is consistent, though. Planets come in all different masses, densities, and luminosities. Similarly, galaxies can vary widely in size due to their stellar content and composition.

Galaxies on the smaller side of life are called *dwarf galaxies*. They may consist of only several million (only!) stars. Larger galaxies can host hundreds of trillions of stars. Because it would be impossible to individually study each of the millions or billions of stars that make up a galaxy, scientists generalize their behavior (and therefore that of the galaxy) through the field of spectral synthesis.

TIP

Studying the radiation and luminosity of a galaxy's individual parts (stars, dust, and gas — hello, old friends!) provides fantastic clues about the age and history of a galaxy. For example, spectral synthesis allows scientists to measure lines of radiative emission from the main three parts of a galaxy, in turn helping us understand a galaxy's overall brightness and age.

The following is a short list of possible galactic radiation sources:

>> **Dust:** thermal emission, light scattering, absorption

>> **Gas:** molecular clouds (cold), shocked gas (warmer), x-ray gas (hot)

>> **Stars:** photometric emission, stellar winds, accretion

The overall brightness of a galaxy is the sum of the electromagnetic radiation emitted by all these processes, over a huge range of wavelengths. Because different processes dominate at different wavelengths, though, a graph of the spectrum of a galaxy across all these different wavelengths would show peaks from the individual components emitting radiation. Scientists use those peaks to help determine a galaxy's age and level of activity.

TECHNICAL
STUFF

Here is another example. If a galaxy has more stars that are red than blue, these red stars are relatively cooler and older (see the Main Sequence in Chapter 5); predominantly blue stars indicate newer, hotter stellar activity. Astronomers can then collectively use this information to help determine a galaxy's luminosity and age. Another interesting parameter is called the luminosity function, a description of the relative number of galaxies with different luminosities (overall brightness) in a particular volume of the universe. Table 10-2 has a comparison of the contents of a few relatively nearby galaxies.

TABLE 10-2   **The Contents of a Few Better-Known Galaxies**

Galaxy Name	Approximate Number of Stars	Home Constellation	Galaxy Type	Catalog ID	Distance from Earth	Diameter
Milky Way	Hundreds of billions	YOU ARE HERE	Barred spiral	M24 (sort of)	YOU ARE HERE	100,000 light-years
Andromeda	1 trillion	Andromeda	Spiral	NGC 224; M31	2.5 million light-years	220,000 light-years
Cygnus A	Billions	Cygnus	Elliptical (radio galaxy)	NGC 6910	760 million light-years	depends on wavelength!
Large and Small Magellanic Clouds	Large: 30 billion; Small: several hundred million	Large: Mensa, Small: Dorado	Irregular, satellite galaxies	Large: NGC 1854; Small NGC 292	Large: 160,000 light-years; Small: 190,000 light-years	Large: 14,000 light-years; Small: 7,000 light-years
Virgo A	1 trillion	Virgo Cluster	Elliptical; radio galaxy	NGC 4486; M87	54 million light-years	Visible light: 170,000 light-years

Excuse you! Emissions of active and inactive galaxies

Some galaxies are considered *active galaxies* (see Chapter 9). These meet a range of criteria, including containing an AGN (active galactic nucleus) at the core.

TECHNICAL STUFF

The AGN is an extraordinarily luminous core area found in active galaxies; it's bright enough that it emits radiation spanning the full electromagnetic spectrum. Active galaxies emit radiation in all wavelengths, from shorter wavelengths (gamma rays) up through the longest radio wavelengths.

Not all of us run a marathon, swim 50 miles, and then head to the gym to finish the day with a light workout. Although around 10 percent of known galaxies are active, the remaining 90 percent are more sedentary, less-than-active galaxies. Don't judge them, though, because even so-called "inactive galaxies" give off electromagnetic radiation. Stars emit light energy and, in a "regular" or inactive galaxy, it's thought that the total energy emitted from that galaxy is an additive sum of the energy released from every constituent star. This energy then spreads out in all directions and gives the galaxies their bright appearance in our night sky.

REMEMBER

Active galaxies, on the other hand, have two sources of energy emission: the same star emission energy as found in all galaxies, plus the energy emitted from its active galactic nucleus. Recall our bright-and-fast friends, quasars and blazars, from Chapter 8 — these two AGN are bright sources of radio emission that are thought to be caused by material falling toward a supermassive black hole at the center of a galaxy. Watch out, though — there are other kinds of active galaxies out there!

Both large and active: Seyfert and radio galaxies

Of the main types of active galaxies, two shining examples (that'll make sense in a minute) are radio galaxies and Seyfert galaxies.

REMEMBER

Radio waves are a longer-wavelength form of electromagnetic radiation, but one that is easy to detect with radio telescopes here on Earth. Both radio and Seyfert galaxies are powered by the supermassive black holes at their cores.

Radio galaxies are a kind of galaxy, usually elliptical, that contain strong AGN. They earn their name by emitting most of their radiation from the radio wavelength portion of the EM spectrum, typically in the form of narrow jets that feed material out into a pair of huge, bright lobes visible at radio wavelengths. These lobes extend far beyond the edges of the visible galaxy. You can't see the radio lobes because radio waves aren't visible to the human eye, but they are detectable with radio telescopes. In fact, radio galaxies are the strongest source of electromagnetic radiation in the radio portion of the spectrum. They also tend to give off x-rays from their cores. Figure 10-3 shows the radio galaxy Hercules A in visible images from the Hubble Space Telescope and radio images from the Very Large Array. Note the huge jets feeding lobes that are far larger than the central galaxy.

FIGURE 10-3: Combined image of radio galaxy Hercules A, showing the central galaxy in a visible Hubble image and the radio lobes in a radio-wavelength observation from the VLA.

Courtesy of NASA, ESA, S. Baum and C. O'Dea (RIT), R. Perley and W. Cotton (NRAO/AUI/NSF), and the Hubble Heritage Team (STScI/AURA)

TECHNICAL STUFF

Another type of galaxy with AGN, the Seyfert galaxy (named after American astronomer Carl Seyfert), emits very strong waves of infrared radiation. Seyfert galaxies are typically spiral and tend to be very luminous. Most look like normal galaxies in visible wavelengths, but infrared images show their intense emissions. As opposed to radio galaxies, most Seyfert galaxies don't give off significant energy in radio wavelengths, though some have radio jets. Figure 10-4 shows Seyfert galaxy NGC 1566 as seen in infrared wavelengths. The central core of this image is extremely bright and is diagnostic of a Seyfert galaxy. There are also star formation regions visible in its two spiral arms.

Black Holes and Their Role in Galaxies

What keeps a galaxy from disintegrating back into the primordial ether from whence it came? Gravity, that's what. Gravity is the tie that binds, and a force to be reckoned with.

Black holes, the doughnut holes of the galaxy

REMEMBER

To their best knowledge, scientists believe that the vast majority of galaxies have a supermassive black hole at their core. How massive is supermassive? Black holes are traditionally categorized according to their mass:

>> **Primordial black holes:** Theoretical, thought to range from 0–100,000 solar masses

>> **Stellar-mass black holes:** Formed from collapsed neutron stars, about 3–100 solar masses

» **Intermediate mass black holes:** Thought to range from about 100–100,000 solar masses

» **Supermassive black holes:** At center of most large galaxies, >100,000 solar masses

It's thought that supermassive black holes grow in size along with the size of the galaxy. Doing what they do best, supermassive black holes pull in surrounding dust and gas at a pace that would make your household vacuum cleaner go on strike. Excess material that hasn't yet been absorbed into the black hole forms an accretion disk. The brilliant x-ray energy given off here results in the quasars and blazars described in Chapter 8, as well as radio galaxies and Seyfert galaxies. Many galaxies seem to glow with x-ray radiation produced by this material falling toward their central supermassive black hole.

All galaxies are surrounded by a halo of dark matter, a (as yet theoretical) ring that serves as a type of frame for the visible matter seen in a galaxy.

TIP

Dark matter is, by definition, not luminous; it's invisible material that scientists believe exists in space but can't be seen because it doesn't emit light. It's thought that these dark matter halos provide a path by which a galaxy's central black hole begins forming. Because you can't observe dark matter directly, you can only look for its indirect effects.

REMEMBER

Dark matter halos were detected by studying the rotation of galaxies. It turns out that the inner and outer portions of a galaxy all rotate at similar rates, requiring extra mass on the outside of the galaxy — much more than is observed based on its brightness. The presence of non-light-emitting or absorbing mass, or dark matter, in a halo surrounding a galaxy or galaxy cluster solves this problem. More to come in Chapters 11 and 15.

One proposed explanation for the formation of supermassive black holes is the collapse of a dark matter halo, one that produces a seed black hole that grows to galactic size. A seed black hole would have enough initial mass to grow quickly, fast enough to explain the supermassive black holes observed very early in the history of the universe. If such black holes formed through the more conventional method of the collapse of a massive dust cloud, their growth rate wouldn't be fast enough to create supermassive black holes so early in history.

Dynamics and observation

REMEMBER

Black holes haven't been directly observable until very recently. The first image of one was released in 2019 by the Event Horizon Telescope Collaboration (refer to Figure 8-2). Dubbed the "fuzzy orange donut" by astronomers, the first black hole lucky enough to be captured Hollywood-style is a supermassive black hole at the center of the M87 galaxy, and it features a string of light surrounding the proverbial black (donut) hole.

Although you wouldn't want your sugary cronut on a rotisserie, rotation comes into play in two ways with supermassive black holes:

>> If a supermassive black hole was created by pulling in matter with rotational energy, the black hole keeps and amplifies that rotation in its own internal rotation.

>> Galaxies orbit about their center of mass. Because a galaxy's supermassive black hole is at the center, the galaxy incidentally rotates about the black hole.

In addition, consider how a black hole affects the stars in its galaxy. As a case study, the XMM-Newton X-ray Observatory from the European Space Agency (ESA) recently captured data from the accretion disk surrounding the black hole in an active galaxy, PG 1114+445. Results indicated that energy outflow from the black hole's accretion disk actually pushes gas away from the center of the galaxy which, in turn, may make star formation less likely.

REMEMBER

Remember your best buddies, dust and gas? Without them, normal stars do not form. These black hole outflows affect the whole evolution of galaxies; after supermassive black holes form and become massive enough to have substantial outflow jets and luminosity, they clear out much of the remaining gas and dust from the young galaxy. This clearing then causes star formation to slow down and eventually cease. Astronomers think that this theory may explain why more mature galaxies seem to have far fewer new stars forming than younger galaxies.

The Hubble Deep Field

TIP

In 1995, astronomers pointed the Hubble Space Telescope at the constellation Ursa Major, at a dark portion of the sky with no known galaxies or stars. Over the course of 10 days, the telescope took 342 images at multiple wavelengths that were then combined to produce an image of the sky as never seen before. The image revealed hundreds of galaxies, clusters, and other strange astronomical objects that astronomers had never previously seen. The sensitivity of Hubble, orbiting the Earth well outside of the absorption and blurring of the atmosphere, gave astronomers unprecedented access to faint galaxies at high redshifts.

The experiment was repeated in 2004, after astronauts had installed an even more sensitive camera on the Hubble Space Telescope. This time, the "Ultra Deep Field" image was taken on a patch of sky near the constellation Orion. Figure 10-5 shows the Ultra Deep Field image and its collection of nearly 10,000 ancient galaxies. The most ancient of these formed only 800 million years after the Big Bang, whereas the brighter ones are closer and formed about a billion years ago (or 13 billion years after the Big Bang). These galaxies are more irregular in shape than today's galaxies, and it's likely that the process of galaxy formation was higher-energy and more chaotic in ancient galactic history than it is more recently.

The new James Webb Space Telescope is also being used to take deep field images that probe back further in time, to view ancient galaxies at high redshift that date to the early days of the universe. Images such as these can be used to study cosmology — the formation, and ultimate fate, of the universe. How were the very first stars, and galaxies, in the universe formed? Check out Chapter 14 to see how formation processes were similar, but fundamentally different, in the early days of the universe, and don't miss Chapter 16 for a discussion of the end of it all.

FIGURE 10-5:
The Hubble Ultra
Deep Field.

Chapter **11**

Bigger Than Huge: Galaxy Clusters

I f two's company and three's a crowd, what do you call the trillions of galaxies out there? One big party, that's for sure. Any good gathering needs crowd control to keep the masses from stampeding, and that's where galaxy clusters come in. When two galaxies went to a party and were introduced by your friends Dust and Gas, a special bond ensued. Now bring a few thousand of their closest galactic friends into the mix, introduce a good helping of gravitational binding, and presto! You have a galaxy cluster, one of the largest gravitationally-bound structures in the universe.

TECHNICAL STUFF

Galaxies aggregate into clusters, both large and small, based on their properties and proximity to one another. An array of physical characteristics including mass, magnetism, temperature, and luminosity all come into play with the mechanics of clustered galaxies. In this chapter, learn more about galaxy groupings near and far, and see how the data gathered about the Milky Way and its friends translates into a collective understanding of galaxies farther out in the universe.

Making Friends: The Basics of Galaxy Clusters

TIP

Opposites attract, but sometimes there's companionship in familiarity. It's thought that most galaxies exist as part of a galaxy cluster. But what exactly constitutes a cluster, as opposed to a few galaxies that happen to be near each other? The main distinction is gravity. If you have a few galaxies gravitationally bound to each other, you have a galaxy group. More gravitationally bound galaxies make up a galaxy cluster, and multiple galaxy groups and clusters create a galaxy supercluster:

>> **Galaxy groups:** < 100 galaxies

>> **Galaxy clusters:** < Several thousand galaxies

>> **Galaxy superclusters:** 3–10 galaxy clusters

Just like with star clusters, galaxies that appear to be adjacent (as seen from the Earth) may not actually be a gravitationally bound cluster. Figure 11-1 shows a grouping called Stephan's quintet, but only four of these galaxies are a gravitationally bound galaxy group. The galaxy at the left is about 250 million light-years closer to the Earth than the other four, though it appears to be right next to the other galaxies. You can see the four bound members of the group interacting and even pulling bright streamers of material off each other, and there are also visible shock waves and regions of star formation.

FIGURE 11-1:
Stephan's quintet, a galaxy group, as seen in the infrared by the James Webb Space Telescope.

Courtesy of NASA, ESA, CSA, STScI / Public Domain

The ins and outs of a galaxy cluster

TECHNICAL STUFF

Galaxy clusters are made of (who would have thought?) galaxies! But that's not all. Within a cluster, galaxies are separated by luminous, hot gas — and when we say hot, we really mean hot. That intergalactic gas can measure up to 100 million °F (56 million °C) and it's hot enough to emit x-ray radiation. Great news for astronomers, because that means it's observable with x-ray telescopes. The rest of the material in a galaxy cluster is mostly dark matter (there are some rogue stars for seasoning). Because dark matter emits no light, it can't be observed and measured directly, but scientists can infer its existence and quantity based on these three factors:

>> The observable x-ray emission from the cluster's luminous matter

>> The motion of the cluster as a whole

>> Gravitational lensing

Specifically, galaxy clusters are composed of

>> **Dark matter,** invisible but substantial — up to 90 percent of the cluster's mass

>> Clouds of **bright, luminous matter,** typically hot plasma — 10 to 15 percent of the cluster's mass

>> Dust, gas, and stars **(galaxies)** — No less than 5 percent of the cluster's mass.

Although the name *galaxy cluster* has a much catchier sound than *ball of dark matter and plasma with a few stars thrown in,* galaxies do not make up the lion's share of a cluster. That honor belongs to light-bending dark matter, covered in more detail in Chapter 15. It's thought that despite the presence of both galaxies and hot plasma, dark matter provides the gravity required to keep the cluster together.

TIP

Roughly 90 percent of all galaxies are interacting with one or more galaxies in some way. Clusters tend to have more galaxies and higher densities than groups, and their galaxies are more tightly bound.

There are more than 2 trillion galaxies in the visible universe, and most of them are part of some sort of galaxy cluster or other structure. One particularly large galaxy cluster, the Perseus Cluster, was recently imaged by the ESA Euclid Space Telescope, This image, shown in Figure 11-2, reveals about a thousand galaxies that are part of the Perseus Cluster itself, as well as more than 100,000 galaxies that are in the background of this image, some as far as 10 billion light-years away.

FIGURE 11-2:
Perseus Galaxy Cluster, as imaged by the Euclid Space Telescope

Courtesy of ESA/Euclid/Euclid Consortium/NASA, image processing by J.-C. Cuillandre (CEA Paris-Saclay), G. Anselmi, CC BY-SA 3.0 IGO

Mass estimation: Virial, Sunyaev-Zel'dovich, and more

In the field of astronomy, many celestial objects are notoriously difficult to measure because they're too far away from Earth, their existence is inferred rather than observed, they're obscured by other celestial objects — you get the idea. Calculating the mass of a galaxy cluster is no exception because they are largely dominated by dark matter, but astronomers use a type of mathematical inference to help out.

It's possible to measure or estimate several key properties of galaxy clusters; these include

» Distance from Earth to the galaxy cluster

» Light emissions from the constituent galaxies

» Radial velocity of the constituent galaxies

TIP

The *virial theorem*, a connection between average kinetic and gravitational potential energy of objects that are gravitationally bound to each other, allows you to relate mass (hard to determine from a distance) to velocity (much more easily measured). This theorem can be written as

$$m = \frac{v^2 R}{\alpha G}$$

where m is the total mass of the galaxy cluster, v is the average velocity of objects that are in the galaxy cluster, R is the effective radius of the galaxy cluster, α is a factor related to the distribution of mass in the galaxy cluster, and G is the standard gravitational constant. Here is how you can use this relationship to estimate the mass of a galaxy cluster:

1. Measure the radial velocity of galaxies in the cluster.

2. Calculate the radial velocity dispersion, or the statistical variation of those radial velocities around a mean.

3. Create estimates of how the galaxies move within the cluster.

4. Measure how big the cluster is in the sky, also called the *angular size*.

5. Using the distance to the cluster, convert the angular size into the cluster's actual diameter.

6. Finally, plug all these values into the virial theorem to find the total mass of the galaxy cluster.

This method was first used in the 1930s to calculate the mass of nearby galaxy clusters. However (and not for the first time in this book!), something was wacky with these mass measurements. The mass derived using this method is much higher than the estimated mass based on the total brightness of the observable galaxies.

TECHNICAL STUFF

For a typical galaxy cluster, you might find a mass estimate between 10^{14} and 10^{15} times the mass of the sun (a typical unit used for comparisons). However, measuring the brightness of all the galaxies in the cluster, for about 100 to 1000 galaxies, yields a total brightness of 10^{12} times the brightness of the Sun! The galaxy cluster is 200 to 500 times less bright than it should be, based solely on its mass.

German astronomer Fritz Zwicky first noticed this problem when he tried to measure the mass of the Coma Galaxy cluster. He came up with four possible solutions:

>> Stars in the Coma cluster emit much less light than those in the Milky Way.

>> The cluster itself is not in equilibrium, meaning the virial theorem doesn't apply.

>> The laws of physics in the Coma cluster aren't the same as those here in the Milky Way.

>> The Coma cluster contains mass that doesn't give off any light.

Zwicky didn't have a favorite explanation out of these four, but as measurements of other galaxy clusters were made, the same problem arose over and over. Astronomers today think the answer is behind door number 4, and it's called *dark matter.*

Like any good problem, we can confirm that the mass of a galaxy cluster isn't all visible in more than one way. In addition to looking directly at the cluster, we can also look at how the hot gas in the cluster affects background light. Roughly 400,000 years after the Big Bang, our universe cooled to the point that neutral atoms could form, and when this happened, light that had been trapped by all the free electrons made a break for it. Those photons now appear as a wall of microwave radiation shining at us from every direction that we call the *cosmic microwave background* or *CMB* (see Chapter 13). As that light tries (it's very trying) to move through these massive galaxy clusters filled with hot gas, free electrons remind the photons of where they came from by blocking their way. This effect — named after its discoverers, Russian astrophysicists Rashid Sunyaev and Yakov Zel'dovich — means maps of the CMB have little gaps where hot gas in galaxy clusters steals the microwave light. The size of these gaps can be used to estimate the mass of a galaxy cluster.

The Sunyaev-Zel'dovich effect can also be used to detect very faraway galaxy clusters at very high redshift, because the presence of this effect means there has to be a galaxy cluster, even if it's too far away to resolve individual galaxies. The European Space Agency's Planck space telescope used this technique successfully.

Galaxy distribution throughout space

As if galaxy clusters weren't big enough, how about something even larger? A galaxy supercluster contains other galaxy clusters and groups. The Laniakea supercluster, for example, is a supercluster of about 100,000 galaxies, including our own Milky Way. At least we're in good company!

If you're looking for even larger celestial clusters, welcome to filament territory. Some of the largest structures in our universe seem to be filaments, or walls of millions of galaxy groups and clusters that surround empty spaces called voids (spaces without galaxies). This large-scale structure is evolving with time, and we can see the voids become emptier and the walls and filaments grow more defined as we look at images that show us the universe at different ages.

Galaxy filaments can be hundreds of millions of light-years long, but only 20 or so million light-years thick. Superclusters are located at the intersections of galaxy filaments. They can be modeled using supercomputers. Astrophysicists currently think that superclusters and filaments are related to primordial fluctuations in density following the Big Bang, ones that produced a slight nonuniform distribution of matter very early in the universe. This matter was then preserved as larger scale structures developed. For more on this topic, see Chapter 14.

A Galaxy Cluster of Our Very Own: The Local Group

Bring it in, folks — it's time to get up close and personal with the galaxy group nearest and dearest to our collective hearts. The Milky Way Galaxy resides in a galaxy group called the *Local Group*. It's one of the smaller groups with less than 100 galaxies and it's certainly getting on in its years — all 13.6 billion of them — but it's home. And because there's no place like home, let's be introspective for a moment and ponder the wonder that is the Local Group.

Location, location, location

First noted by Edwin Hubble in 1936, the Local Group galaxies are a stone's throw from Earth. Or they would be if you could throw a stone 5 million light-years.

TECHNICAL STUFF

The overall diameter of the Local Group is around 10 million light-years, putting it on the smaller end of galaxy groups. Just as Galileo discovered in the 17th century that the Earth orbited the Sun and was not, in fact, the center of the universe, the Milky Way Galaxy is not at the literal center of the Local Group. The gravitational center of the Local Group is located in between the Milky Way and the Andromeda Galaxy.

TIP

The Local Group consists of three large galaxies and dozens of smaller satellite and dwarf galaxies. They include

>> **Milky Way** — Our own home galaxy.

>> Surrounding the Milky Way are satellite galaxies including the Large Magellanic Cloud, the Small Magellanic Cloud, Sagittarius Dwarf Galaxy, and many others.

>> **Andromeda Galaxy** — The closest large galaxy to us, bigger than the Milky Way.

>> Surrounding the Andromeda Galaxy are a number of satellite galaxies including M32, M110, NGC 185, and many more dwarf galaxies.

>> **Triangulum galaxy** (M33) — The third largest galaxy in the Local Group.

>> Some consider Triangulum a satellite galaxy of the Andromeda Galaxy; it has its own satellite galaxies including Pisces Dwarf and other small dwarf galaxies.

>> 10 or more smaller **dwarf galaxies,** which are not satellite galaxies; these include IC10, Phoenix Dwarf, Leo A, and Pegasus Dwarf Irregular.

Constituent galaxies

Using the number of stars as a metric, the Andromeda Galaxy (M31) in the Local Group wins first place with over 1 trillion stars. The Milky Way comes in second with a hundred billion stars, and the third-largest galaxy in our cluster, the Triangulum Galaxy (M33), has a not-so-paltry 40 billion stars. The size of each of the Big Three galaxies has a similar correlation:

>> **Andromeda Galaxy:** Barred spiral galaxy with a diameter of approximately 220,000 light-years

>> **Milky Way Galaxy:** Barred spiral galaxy with a diameter of approximately 100,000 light-years

>> **Triangulum Galaxy:** Spiral galaxy with a diameter of approximately 61,000 light-years

REMEMBER

The remaining galaxies in our Local Group are called *dwarf galaxies.* As we discuss in Chapter 10, dwarf galaxies are smaller in scale, and contain fewer stars than larger galaxies. The largest two of these, the Small and Large Magellanic Clouds, are irregular dwarf galaxies only visible from the Southern Hemisphere. As with all galaxies in a group or cluster, they're gravitationally bound to our Milky Way Galaxy. As Figure 11-3 shows, the galaxies of the Local Group are a bit clustered, with the Small and Large Magellanic Clouds located the closest to the Milky Way.

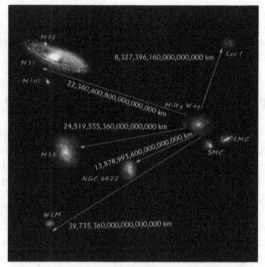

FIGURE 11-3:
Map of the Local
Group with
distances.

Galaxy Cluster Structure and Formation

Rome wasn't built in a day, and neither are galaxy clusters (or almost anything you see in the night sky, for that matter). Galaxy clusters are a tripartite system made up of dark matter, hot plasma gas, and star-galaxy systems. They also comprise the largest celestial objects to be bound together by a gravity of their very own. The formation process begins before even the first stars light up.

Imagine a sort of cosmic speed-dating. Dark matter takes a seat. And waits, and then waits some more. Over time, gas falls in and flirts a little with becoming something, but it will be a few hundred million years before enough mass gathers that stars can turn on, galaxies can start to shine, and things can really start to spark. One galaxy will collect another . . . and another . . . and soon, our dark matter has found the galaxy cluster of its gravitational dreams.

Things don't get too hot too fast, however. That will take time. Over the eons, galaxies interact and lose dust and gas into intergalactic space (along with a few stars for spice). As this gas builds up, it heats up until eventually there is enough hot gas with enough free electrons that the CMB starts getting blocked from interfering (remember that Sunyaev-Zel'dovich effect? We knew you did!).

Although small galaxy groups and clusters are made of mostly dark matter along with the normal galactic ingredient list of stars, gas, and dust, we start to see more and more plasma in larger and larger clusters.

The ever-important role of gravity

Give us a G! Give us an R! Give us an A . . . We think you can see where this is going. Gravity, the force responsible for attracting objects toward each other's centers, is responsible for your not flying off the surface of the Earth. It's the means by which celestial bodies remain in orbit around each other, and is the force that keeps galaxy clusters from sailing off into the celestial sunset.

Because galaxy clusters are the largest gravitationally bound structures in the universe, they can be used to measure the properties of the universe on the largest possible scale. They are dominated by dark matter and are observable at vast distances, so you can also use them to observe the changing characteristics of the universe over time.

Remember that when you look at a far-away object (like a galaxy cluster), you are looking hundreds of millions of years, or even billions of years, into the past because that's how long the light from these galaxy clusters took to reach the Earth. Astrophysicists count the number of galaxy clusters in different mass ranges, and study how they are distributed in space and in time. This helps show how clusters can grow and change over the whole age of the universe.

Astronomers use x-ray telescopes in space to map out hundreds of thousands of large galaxy clusters rapidly. They can combine these observations with optical-telescope measurements of the brightness and spectra of some of the hundreds of galaxies in the clusters. The optical measurements of the spectra let scientists estimate the distance to the galaxy clusters by using their redshift, and also provide another estimate of the clusters' mass. Studying the number of clusters sorted by their mass and their redshift provides a glimpse into how the average size of galaxy clusters changed over time from the early days of the universe.

X-rays and the ICM (Intracluster medium)

The largest percentage of material in a galaxy cluster is, by far, the 80 to 95 percent dark matter. The stars and physical structure of galaxies comprise sometimes as little as 5 percent. In large clusters (and more and more clusters over time), interactions between galaxies can strip gas and dust (and those spicy stars) into the space between the galaxies, the *intracluster medium* (*ICM*) as well, and gravity smashes it together and makes it shine in x-rays. Hot plasma is an ionized gas, so hot that the electrons have been stripped off the nuclei of the mostly-hydrogen gas molecules. The ICM plasma is critically important in astrophysics.

The ICM gets hotter as it gets denser because, put simply, gas does not like to be compressed, either in a can or in a cluster. As gravity smashes things together and gets the particles orbiting fast, those gas rules you may have learned in high school drive the temperature up. Due to this high temperature, the ICM gives off x rays through a process called *Bremsstrahlung emission*, and the brightness of the x-ray radiation is proportional to the square of the density of the ICM.

The ICM is very tenuous, though, and can have a density as low as 10 to 10,000 particles per cubic meter of the galaxy cluster. Note that the density does increase as you move toward the center of the cluster. Th x-ray emission from the ICM is one of the best ways to detect and study galaxy clusters.

The physics of the ICM is still not well understood by scientists. Here's one example: The active supermassive black hole that is often found at the center of galaxy clusters seems to keep the ICM from cooling near the galaxy cluster's center. As material is drawn toward the black hole, high-energy jets of material can be sent back into the ICM. Upon arrival, they seem to lead to the inflation of bubbles or cavities of material and result in the heating of the ICM gas.

Figure 11-4 shows an image of galaxy group NGC 5813, as observed by the Chandra X-ray Observatory satellite. This x-ray image shows multiple cavities in the ICM surrounding the central galaxy in the cluster, thought to be due to eruptions from the supermassive black hole in that galaxy. It's thought that these cavernous spaces may actually slow down star formation because they push out dust and gas, and spread it in ways that make it inhospitable to star formation.

FIGURE 11-4: Interaction of a supermassive black hole at the center of a galaxy group with the ICM, as seen by the Chandra X-ray Observatory.

Courtesy of x-ray: NASA/CXC/SAO/S.Randall et al., Optical: SDSS

The ICM also serves as a repository for the heavy elements produced by the stars during their lifetimes in the galaxies that make up the galaxy cluster. These elements, including oxygen, iron, and silicon, may build up in the ICM until they eventually are incorporated into new stars or galaxies.

Physics of Galaxy Clusters

TECHNICAL STUFF

Galaxy clusters are what's known as *closed systems*. In the world of physics, particularly relevant in thermodynamics, an *open system* is one where energy, heat, and matter flow in and out of a system's environment. A closed system is one where that information stays within the system. Imagine a steaming cup of coffee releasing heat into the world — that's an open system. Now place a large, dense cookie over the top of that cup to contain the steam, and we've created more of a closed system.

If the take-away here is that galaxy clusters are surrounded by cookies, it's probably lunchtime. Galaxies are not considered closed systems because whenever one of their stars goes through a supernova explosion, gas is lost from the galaxy. A galaxy cluster, on the other hand, tends to retain even the gas from supernovas in its ICM, and ICM can accumulate the heavier elements created through nucleosynthesis. Astrophysicists can use a galaxy cluster and its chemical composition to track the history of nucleosynthesis over the age of the cluster.

The unique interactions of dark matter with baryons

REMEMBER

Putting aside actual stars and the gas in the ICM for a minute, we can look at that other 80 to 95 percent of a galaxy cluster — dark matter. It's invisible, even to our telescopes and cameras, because it neither absorbs nor reflects energy. What is dark matter made from? First, consider what non-dark matter is made of. Bodies in the universe are made from matter, which is made from atoms; these are the subatomic particles called protons (positive charge), neutrons (no charge), and electrons (negative charge). See Chapter 2 for more on this. Collectively these subatomic particles are called *baryons*, and matter made of baryons is called *baryonic matter*.

Baryonic matter is visible because it can absorb, emit, and reflect light, or electromagnetic radiation. Dark matter, by definition, does not. A major question in the study of galaxy clusters is whether that dark matter is baryonic or non-baryonic (containing particles other than baryons), and the current thinking is that dark matter is primarily non-baryonic.

TECHNICAL STUFF

If dark matter was cold, the particle velocities would be non-relativistic (small compared to the speed of light). In this scenario, you'd expect the distribution of galaxy clusters to have sharp boundaries; if the dark matter was hot, particle velocities would be relativistic (significant fractions of the speed of light) and you'd expect the distribution of galaxy clusters to be smooth. In fact, the structure of galaxy clusters and superclusters is somewhere in between, suggesting that an additional particle is needed to explain dark matter.

At the end of the day, what is dark matter? One suggestion is that it could be WIMPS — "weakly interacting massive particles." Scientists think that these particles could be anywhere from 10 to 100 times as massive as a proton, but they only have very weak interactions with other kinds of matter. These interactions make them very difficult to detect. One example of these types of particles, the neutralino, is theoretical and has yet to be detected, though gamma ray and neutrino telescopes may be able to provide indirect detections. Direct detection might require specialized Cryogenic Dark Matter Search experiments, which are conducted deep underground using specialized equipment.

Another possible source of dark matter are sterile neutrinos. Neutrinos are very small particles, so small as to be barely detectable, but they're also extremely common. Neutrinos are out there balancing energy-releasing reactions far and wide, helping the Sun fuse hydrogen and your local nuclear reactor keep the power flowing. Neutrinos are harmless, and 100 trillion pass through your body unnoticed every second. Sterile neutrinos are a particle that only interacts with active (normal) neutrinos and would have been produced in the early Universe.

The photons given off as these sterile neutrinos decay could potentially be detected with sensitive x-ray or gamma-ray telescopes. Neutrino telescopes such as Ice-Cube (see Chapter 8) can also search for evidence of sterile neutrinos and dark matter annihilation in galaxy clusters. However, neutrinos alone — no matter how numerous — aren't able to explain the distribution of galaxy clusters.

A third recent suggestion for dark matter involves axions. This hypothetical particle type has yet to be detected, but if it exists, it could help explain why matter seems to be more common than antimatter both on Earth and in astrophysical observations. Axions are theorized to be low-mass particles that are spread across the universe in a cold, coherent, wave-like state. New observations and experiments may soon be able to prove, or disprove, this axion theory. Head over to Chapter 15 for more on dark matter.

Shake it up: Gravitational perturbations and shock waves

TECHNICAL STUFF

What about motion in galaxy clusters? For a moment, return to the motion of the solar system. All planets travel in a perfectly elliptical orbit around the Sun, right? Not quite. Newton's law of universal gravitation tells you that gravitational force relates to a product of the two masses, combined with the gravitational constant, divided by the squared distance between them:

$$F = \frac{Gm_1m_2}{r^2}$$

where F is the force, m_1 and m_2 are the two masses, r is the distance between them, and G is the gravitational constant. (If this isn't sounding familiar, check out Chapter 2.)

TIP

Similarly, Newton's third law of motion reminds you that every force has an equal and opposite reactionary force. Now combine these two ideas, and take another look at a planet's orbit around the Sun. As Uranus orbits the Sun, for example, the Sun's gravitational pull keeps Uranus in orbit because there are no competing gravitational forces to drag it out of its elliptical path. However, per Newton, Uranus exerts a gravitational force on the Sun in return, as do the other planets orbiting the Sun. The total effects of these small gravitational pushbacks are called gravitational perturbations, and can be expressed as:

$$F_{total} = F_{Sun}(1 + \frac{\Delta F}{F_{Sun}})$$

where F_{total} is the total gravitational force, F_{Sun} is the force due to the Sun, and ΔF is the gravitational perturbation. (All right, that was a bit simplified — all the planets pull on each other, too, and this weird truth allowed researchers in the 19th century to find Neptune because it tugged Uranus around a bit.)

Now, bring this knowledge back to galaxy clusters. Suppose that a really large galaxy cluster has a thermodynamically stable configuration of gas. Within a galaxy cluster, the stability of the gas can be challenged by gas accretion within the ICM. As gas accretes into clumps, the act of accretion causes gravitational perturbations within the gas cloud which, in turn, induce turbulence.

This turbulence leads to shock waves, or waves of pressure through matter that change its properties. In turn, these shock waves alter the thermodynamics of the gas cloud and can accelerate particles to close to the speed of light, at relativistic velocities. This particle acceleration also produces visible filaments of particles which follow giant magnetic field lines surrounding the cluster, and the energy from this shock wave also gives off a huge burst of energy detectable by radio telescopes from Earth.

Always under pressure: Radiative gas physics

Gravitational perturbation and shock waves show different ways in which scientists can better understand the physics of galaxy clusters. Before adding a third method into the mix, we need to take a closer look at the thermometer. We start with defining a few terms:

>> **Radiative cooling:** The thermal process by which a body loses heat by radiating away photons (light or radiation). Example: the way the ground cools off in the evening after the Sun has faded over the horizon.

>> **Radiative heating / thermal radiation:** The thermal transfer of heat energy via radiation. Example: the warming of the air around a campfire.

>> **Relativistic energy:** The idea, per Einstein, that energy and mass are two aspects of the same thing. This idea is better known as $E=mc^2$.

Galaxy clusters contain large clouds of hot gas which exhibit radiative cooling over time. This effect is countered by the addition of thermal radiation into a galaxy cluster, and this addition could occur from star formation or the explosion of a supernova. And let's not forget about the supermassive black hole at the center of most galaxies! Its accretion disk provides a continual source of energy. All these various thermal processes inside a galaxy cluster can cause relativistic outflows of energy and momentum (mass and energy accelerate at speeds close to the speed of light). These outflows can potentially spread far and wide — and we're talking thousands of light-years wide — because of their rapid velocity, and these outflows have the potential to affect the stability of the entire galaxy cluster.

REMEMBER

You've got your gravitational disruptions, shock waves, and thermal changes all affecting the physical behavior of galaxy clusters. What's next? Now that you've got a solid handle on the thermal processes of galaxy clusters via radiative gas physics, remember non-thermal processes such as magnetic fields. These can modify the gas turbulence, and can cause particles to become accelerated to relativistic speeds. Scientists can observe this acceleration via radio telescopes.

Galaxy and cluster mergers

Much like a business merger might combine two corporations into a single mega-corporation . . . Actually, galaxy and cluster mergers are nothing like business mergers — except, in a way, they are. Ever heard of a hostile corporate takeover? That's nothing compared to the world of merging galaxies; one galaxy may steal resources from another, and it wouldn't be unheard of for a galaxy to turn cannibal and swallow up neighboring stars. Who knew space could be so cutthroat?

If interacting galaxies merge to form a bigger galaxy, why will multiple galaxies sometimes join together to form a galaxy cluster? Before we talk about galaxy cluster mergers, we start with lower-stakes galaxy mergers. Then we will move on to galaxy cluster mergers, which are a huge deal: When two galaxy clusters merge with each other, they can give off the biggest amount of energy seen in the universe since the Big Bang.

An involuntary story of acquisition: LMC and SMC

Start small by thinking about galaxy mergers — well, as small as you can get when you're talking about some of the biggest structures in the universe! Large galaxies like the Milky Way formed through mergers of smaller galaxies.

TIP

It's thought that up to 25 percent of all moderately sized galaxies are currently merging with other galaxies, making mergers a very common occurrence. When such a merger takes place, there's give and take — emphasis on the taking — as stars near the borders of their respective galaxies may hop to a new parent galaxy. Exchanges of this sort can affect the structure of both galaxies, because one may be losing a star whereas the other gains one.

TECHNICAL STUFF

One good example of galaxy merging can be seen in the Small and Large Magellanic Clouds (SMC and LMC), irregular galaxies within our Local Group that are gravitationally-bound satellite galaxies of the Milky Way. Formed at approximately the same time as the Milky Way (12+ billion years ago), these galaxies are thought to contain streams of stars that were merged in from other galaxies.

During these galaxies' formative periods, evidence suggests that their original parent galaxy may have passed close enough to the Milky Way that its powerful gravitational pull brought the LMC and SMC into the Milky Way's sphere of influence. The net effect here would have been the stealing of stars from these now-satellites of the Milky Way. Alternatively, it's also possible that the SMC was broken apart into pieces from a larger galaxy that interacted with the LMC.

TIP

What happened to the LMC and SMC's original parent galaxies after this despicable act of star-stealing? Have you heard the myth that a mother bird will reject her babies if they've had human contact? (And yes, this is almost entirely a myth.) More than likely, the LMC and SMC parent galaxies did not attempt to chase down their offspring in some gravity-defying burst of cosmic energy, but instead it's much more likely that they simply drifted off in another direction.

Figure 11-5 shows three galaxies in the process of merging into a single, larger galaxy, as seen by the Hubble Space Telescope. This trio is called SDSSCGB 10189. The process of merging is distorting the spiral structures of all three galaxies.

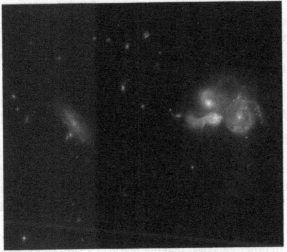

Courtesy of ESA/Hubble & NASA, M. Sun

FIGURE 11-5:
Merging galaxies
as seen from the
Hubble Space
Telescope.

Prediction via computer modeling

If a spherical galaxy merges with another spherical galaxy, would you end up with a couple of tennis balls joined together at the hip? Not so, say scientists who use current observations of galaxies as a basis for advanced computer modeling. Fourth-century BCE philosopher Aristotle was one of the first to posit that a whole body may not simply be a sum total of its parts (and he certainly didn't need a computer to make that statement!). Scientists have used high-powered computers as well as camera choreography and rendering to visualize mergers between galaxies and, in the process, try to explain the form that results when two or more galaxies collide. These simulations have shown that both off-center elliptical and many irregular galaxies may be the result of merged spiral galaxies.

WARNING

Stand back when galaxies are merging! The gravitational interactions between objects as massive as galaxies can be sufficient to trigger a massive burst of star formation. Yes, you read that right — galaxy mergers can actually cause new stars to form. Dubbed *starbursts*, this period of rapid star formation is caused by compression of the gas due to gravitational interactions during the merger, resulting in shock waves that lead to star formation.

Scientists predict that billions of years down the road, the Milky Way will likely collide with the Andromeda Galaxy; what might the resulting merger bring? The Milky Way's spiral arms will be destroyed, and our central supermassive black hole will be consumed by Andromeda's even more supermassive black hole. Some stars will be ejected in this galactic merger, but most will join the new giant galaxy.

Galaxy cluster collisions

REMEMBER

Galaxy clusters have little regard for right of way. Colliding galaxy clusters can merge into larger scale structures than the traditional galaxy. Astronomers have created a theory in which large galaxy clusters are built up into superclusters over time through continual mergers of smaller clusters. As galaxy clusters get close enough to each other to start the process of gravitational attraction, different processes are possible depending on the relative velocities of the two systems. Both high-energy collisions and lower-energy mergers are possible.

TIP

If one galaxy cluster is much larger than the other, the result could be a process of accretion where one cluster simply acquires the other.

In any of these cases, shock waves are formed in the collision at non-relativistic velocities for galaxy cluster collisions. The kinetic energy of these shock waves heats up the plasma in the ICM, but the shocks can also

>> Accelerate particles

>> Amplify magnetic fields

>> Lead to magnetohydrodynamic turbulence in the system

Figure 11-6 shows a galaxy cluster collision, Abell 2146, as seen in a combination of x-ray data from the Chandra Space Telescope and optical data from the Hubble Space Telescope. The bright diffuse areas are regions of hot gas in the two clusters, and the bright central region is a location where gas from one galaxy cluster is pushing through the hot gas from the other cluster. This gas glows brightly in the x-ray Chandra observation, and data from Hubble fills in the visible galaxies and stars.

FIGURE 11-6:
Galaxy cluster collision Abell 2146, as seen in a combination of x-ray and optical data from the Chandra and Hubble space telescopes.

Courtesy of x-ray: NASA/CXC/Univ. of Waterloo/H. Russell et al.; Optical: NASA/STScI

What Galaxy Clusters Tell You About the Universe

Ask an astronomer to identify a constellation and you'll likely get a three-hour lecture about the history of the universe (as they avoid letting on that they can't find constellations without a computer). One of the many reasons astronomers gravitate (ha!) to the field of astronomy is the lure of understanding the world at a fundamental level. What exactly happened at that moment of the Big Bang? What can the Milky Way tell you about the origins of life? And, as always, what clues can you find about life in the universe?

Galaxies are no exception in that they provide astronomers with tantalizing details of fundamental cosmology. It's thought that most galaxies formed and subsequently clustered by the time the universe was about 6.75 billion years old, roughly half its age today.

The implication here is that galaxy clusters initially started as protoclusters, or very young formative galaxies with a high redshift, and that this phase of development was frantically active in its production of stars. What information can you glean about the cosmos here? What can you learn about the fundamental structure and history of the universe from its largest gravitationally bound structures? Quite a lot, as it turns out.

Bigger may be better

Good things often come in small packages. Take the infinitesimally tiny quarks as an example. They build the protons and neutrons that make matter possible, and, without them, none of us would be here. Sometimes, though, size has an advantage. Galaxy clusters are some of the largest objects in the universe. As such, they are a microcosm of the origins of the universe, and they allow you to illustrate the undetectable — you guessed it, dark matter again.

One of the biggest "uses" of galaxy clusters is in helping locate regions where dark matter is more common. If you see a galaxy cluster, you know that you're looking at a region where there's a higher density of dark matter. The more massive a galaxy cluster, the more dark matter it likely has. One analogy is looking out the window of an airplane at night. You can't actually see the cities below you but if there's a higher concentration of lights, you can infer that a city is present. The higher the density of lights, the larger area they cover (and the larger the city). The visible galaxies of a galaxy cluster are like those city lights, indicating the presence of the invisible dark matter that's also there.

The dark matter environment surrounding a galaxy cluster likely influences position — whether the galaxy cluster is more closely packed or is more spread out — as well as how far away it is from other galaxy clusters. Recent observations have shown that galaxy clusters whose galaxies are more closely packed together seem to be farther away from neighboring galaxy clusters than those whose galaxies are more spread out. Figure 11-7 shows Abell 1689, a densely-packed galaxy cluster, as seen by Hubble.

FIGURE 11-7: Abell 1689, a densely-packed galaxy cluster, as seen by Hubble.

Courtesy of NASA/ESA/JPL-Caltech/Yale/CNRS

This observation could indicate that the cluster-forming dark matter environment depends on the cluster's age. Massive older galaxy clusters seem to have a structure and distribution that tracks small quantum fluctuations in the distribution of matter only a fraction of the second after the Big Bang (head to Chapter 13 for more information). Galaxy clusters can then be used to shed light on processes very early in the history of the universe, and they can also help you understand the evolution of large-scale structures.

Chemical emissions as prognosticators

Remember those metals and the idea of the closed system? Because galaxy clusters are large on the cosmic scale, they hold onto all the higher products of nucleosynthesis produced by stars in their galaxies. Accordingly, scientists can use these metals to track the different chemical elements produced over the age of the galaxy cluster which, per the above, can be a sizable fraction of the age of the universe. Chemical abundances are like a fossil record of what happened over the age of the galaxy cluster.

Astrophysicists perform these types of analysis using spectroscopy — including x-ray spectroscopy, because that's the wavelength that these distant high-energy ICM molecules give off. X-ray telescopes like Chandra, XMM_Newton, and Suzaku observatories have detected emissions from elements such as oxygen, neon, magnesium, silicon, sulfur, argon, calcium, iron, and nickel from galaxy cluster ICM measurements, and carbon and nitrogen can be detected for closer galaxy clusters.

You can use these measurements not only to detect different chemical elements, but also to study their distribution within the galaxy cluster. The ICM is the general repository of all this material, and models suggest that it could be transported from its formational stars out into the ICM through interactions with plasma.

REMEMBER

In one model, the relatively cool metal-rich gas (for astrophysicists, "metal" actually means anything heavier than hydrogen) is uplifted by bubbles of relativistic plasma created by the jets of the AGN; this is the supermassive black hole at the center of the galaxy cluster. These plasma bubbles spread the metals out into the ICM, between the galaxies of the cluster, where it stays in a relatively stable environment. The plasma bubbles emit radio waves that can be detected by radio telescopes to help further study this process.

Change is hard: Slow rates of change in galaxy clusters

Protoclusters, or very early star-forming galaxy clusters, explode into a burst of star formation. Beyond this initial birthing period, though, galaxies are not known for being speed demons. Most galaxies have been shown to create stars and burn through their gas supplies over billions of years, though bursts of star formation can arise during galaxy mergers or when a galaxy falls into a cluster (or other gravitational interactions). The advantage of this slower rate of change is the imprints left behind, ones that leave a breadcrumb trail of their formation history.

Because galaxy clusters seem to be in equilibrium, they are generally stable over long periods of time, and their structure seems to preserve some of the initial aspects of the distribution of matter in the very early universe. Galaxy clusters with massive sizes can be observed back to very high redshift, implying that some of these structures formed very early in the history of the universe. They can also be compared to those structures that you see at lower redshifts, meaning ones that formed later. This general stability of galaxy clusters allows scientists to make inferences about their changes over time.

Figure 11-8 shows an ancient galaxy cluster, SPT2215, one that's located about 8.4 billion light-years from Earth. The light you see from it was given off when the universe was only about 5.3 billion years old, less than half its current age. In this image, Hubble data shows the individual galaxies, including the large bright central galaxy, and Chandra x-ray observations illuminate the bright diffuse ICM between the galaxies.

FIGURE 11-8:
Ancient galaxy cluster SPT2215, as seen in a combination of images from the Chandra X-ray Observatory and the Hubble Space Telescope.

Courtesy of x-ray: NASA/CXC/MIT/M. Calzadilla; UV/Optical/Near-IR/IR: NASA/STScI/HST; Image processing: N. Wolk

TIP

SPT2215 is particularly interesting because it has a very regular, relaxed structure as seen in its distribution of gas, and there's no indication of violent mergers with other galaxy clusters. Astrophysicists had predicted that this kind of ancient galaxy cluster formed at a time when the universe was smaller (because it's been expanding since formation), and interactions between galaxy clusters would have been frequent; it's therefore quite rare to observe a cluster that appears to have been left alone for at least a billion years.

In addition to all their other interesting properties, therefore, galaxy clusters also may preserve some primordial structure from the very early days of the universe, just after the Big Bang. See Chapter 14 for more on how those first galaxies might have formed.

Chapter **12**

Weird and Wacky Galactic Phenomena

A s the old saying goes, things are not always as they seem. Ancient Greek astronomers believed we lived in a geocentric universe with the Earth at the center. This belief was prevalent until the 16th century, when Copernicus first knocked us earthlings down a peg with the heliocentric theory of all bodies orbiting the Sun. Not to disagree with our ancestors — well, maybe just a little — but advanced science combined with an inquisitive mind opened the door to a new and deeper understanding of the cosmos.

Here are a few updates to your existing knowledge of physics in galaxies:

» Light always travels in a straight path, right? **Wrong.**

» Can you see dark matter with a special telescope? **Nope.**

» Wormholes only exist in movies, don't they? **Maybe not.**

The strangeness of galaxies provides us with many excellent opportunities to explore astrophysics. Could you travel through a wormhole, or even back in time? Feeling like dusting off your skills to become a space archeologist? You've come to the right place, so let's dive in!

Not Quite Dinosaurs: Galactic Archaeology

If you ever wanted to dig up Egyptian pyramids as a child, the odds are excellent that you and archaeology have crossed paths. Terrestrial archaeology, the kind that's Earth-based, is the study of human history through physically excavating and analyzing artifacts. These artifacts can be bones (human or other), or they can also be arrowheads, pottery, or any objects left behind by humans at some point in history. Incredible knowledge about a society's history, structure, and culture can be gained through terrestrial archaeology.

Astronomers who seek this type of knowledge about galaxies follow a conceptually similar path with galactic archaeology, a field of study devoted to using the Milky Way's celestial artifacts to learn more about its history and makeup. Terrestrial archaeologists have a set of tools — axes, shovels, buckets, brushes — that are used to perform careful excavations. Galactic archaeologists' tools of the trade lie mainly in data; cameras, telescopes and spacecraft are all used to retrieve critical data about the Milky Way that helps astronomers paint a picture of how our galaxy formed, and what events took place during its early years.

TIP

Just as terrestrial archaeology uses artifacts to gain knowledge and reconstruct details about a particular civilization, so does galactic archaeology! Examples of celestial artifacts from the Milky Way might include stars, gas, and data (either velocity or positional).

Without proper technique, Earth-based archaeologists could easily end up doing more harm than good. They have to figure out the right places to dig, they must dig in such a way that they don't damage unseen artifacts, they need to be careful not to break objects as they're being excavated, and all steps must be photographed and cataloged along the way. Galactic archaeologists also need to create or utilize certain techniques when tracing Milky Way history. For example

>> Isotope dating tells us that the Milky Way is about 13.6 billion years old — that's not far from the age of the universe, 13.8 billion years — suggesting that the Milky Way came into being not long after the Big Bang.

>> Analyzing the chemical makeup of stars via spectroscopy tells us how a star's gas and metal content have changed over time, critical information in pinpointing stellar evolution (from star to supernova!).

>> Astroseismology studies a star's oscillations in brightness in order to glean information about a star's age and density.

What little stars are made of

Much like terrestrial archaeologists create a fossil record to catalog and date artifacts, galactic archaeologists study some of the oldest structures in our galaxy, including stars that are very low in metals. The rarest of these stars formed from the remains of the first generation of stars and the leftovers from the Big Bang (see Chapter 14 for more on stellar populations.)

Scientists also study the origin and properties of the Milky Way's globular clusters, groupings of very old stars thought to have formed only a few billion years post-Big Bang. Spectroscopy can be used to measure the chemical makeup of stars in globular clusters as compared to the elements found in more modern stars. This technique can help determine what conditions were like back in the early universe. Since all the stars in each globular cluster were born at the same time and out of the same material, they also are unique laboratories for studying how stars of different masses evolve.

The composition of stars can also be used to track stellar motion throughout the galaxy. Astrophysicists think that each star-forming region in our galaxy — which serve as nurseries for open clusters and the stars they leave behind — has its own unique chemical signature. Studying hundreds of thousands of stars using spectroscopy may allow scientists to figure out which stars formed in the same regions. Even if those stars are now spread around the galaxy, these fingerprints can be used to trace their motions through time and figure out where they were born and how they have traveled. Think of the family reunions!

Stellar ages and astroseismology

REMEMBER

Estimating the age of a star can be quite difficult — it's not like stars come with a "best if used by . . ." date! You can estimate a star's age using stellar evolution — is the star burning hydrogen into helium, or has it moved on to higher levels of nucleosynthesis? (Check out Chapter 5 for a refresher.) You can do this using the H-R diagram to plot the brightness of a star against its color, and this information can be used to figure out a star's position on a plot of its lifetime.

The problem with this approach is that it only works for a limited set of circumstances, and it's not very accurate. Spectroscopy of stars can measure only the chemical makeup of a star's surface layers, but these layers obscure what's going on below. The H-R diagram method gives you a sense of the relative life path of a star but isn't very useful for determining an absolute age.

Another method, called astroseismology, has turned out to be helpful in dating stars. Astroseismology was first developed to study the interior layers of our Sun, and quickly turned out to be a useful method to study other stars as well. Convection on the surface of a star results in pressure changes deep in a star's interior, producing stable waves in a resonance. The frequency of these waves depends on the size of the star as well as what the star is made of. As a star ages and burns more of its hydrogen into helium, the waves can't move as quickly and the frequency of the resonance changes.

TIP

Compare this scenario to a musical instrument. If you pluck a string on a violin, the entire instrument vibrates to produce a particular note. Moving your finger on the string makes that vibrating area longer or shorter, thus altering the resonant frequency and making the note's pitch sound lower or higher.

How can you measure these waves? It turns out that waves make the star expand and contract slightly, causing minor periodic changes in the star's brightness. The changes are faint but measurable, about 0.1 percent of the star's brightness, and can be measured by a telescope. These brightness changes can then be compared with theoretical models that provide the star's mass and its age.

REMEMBER

The Kepler space telescope, launched to look for exoplanets (described in detail in Chapter 7), monitored the brightness of about 500,000 stars in the Milky Way with very high precision. Although this dataset was primarily used to find planets orbiting these stars, it turned out to be very important for astroseismology as well. Astroseismology can also be used to date red giant stars, ones so bright that they can be seen and measured at much farther distances. Kepler and a follow-up mission called the Transiting Exoplanet Survey Satellite (TESS) have allowed galactic archaeologists to figure out the ages of about 100,000 stars, and most of these have turned out to be red giants.

TIP

The Large and Small Magellanic Clouds (LMC, SMC) also provide great fodder for studying galactic archaeology. Astroseismology can be used to study red giants in the LMC, but this nearby satellite galaxy is about as far out as the technique can be used — any farther, and the stars are too dim to measure accurately enough to catch these variations.

Tracing galactic mergers

In addition to knowing how old various stars in the Milky Way are, to reconstruct the galaxy's history you also need to know where the stars are located, where they came from, and where they're going. The ESA Gaia space telescope, launched in 2013, has tracked the positions of hundreds of millions of stars so far, as well how fast they're moving and in what direction. Astrophysicists have been able to combine this data with stellar age measurements to start piecing together the history of the Milky Way, looking for places where the ages or star motions are different from elsewhere in the galaxy.

Galactic archaeology can also be used to search for traces of galaxies that were subsumed into the Milky Way long ago; the method here involves looking at substructures and stellar streams that remain and provide clues to past accretion. Astrophysicists have determined that today's Milky Way was formed from a series of collisions with other galaxies, and are working to try to come up with a chronology of those events and how they resulted in our galaxy.

TIP

For example, galactic archaeologists recently discovered a region of stars in a cylindrical shape nicknamed the Gaia Sausage (named because the shape of this region resembles a sausage). This region may have originally been its own separate dwarf galaxy because its stars all have slightly different chemical compositions, as well as different velocities and orbits, from the rest of the Milky Way. It looks like that about 8 to 11 billion years ago, this dwarf galaxy merged with the Milky Way, bringing about 50 billion solar masses along with it. It's also thought that this merger helped create the thick disk of our galaxy.

High Energy Astrophysics

Astrophysics is all about studying celestial objects that are very big and very far away. Unfortunately for humans, it's a fundamental property of electromagnetic radiation that its intensity decreases with the distance to the 2nd power (this is the inverse square law; see Chapter 2). The implication? Those interesting faraway objects are also very faint. One way around this dilemma is to look for events or objects that give off particularly high-energy radiation, and we can detect this radiation over vast cosmic distances.

Here's how high energy astrophysics fts into the astrophysical puzzle:

>> **Astronomy =** the study of space and celestial objects

>> **Particle physics =** the study of subatomic particles, nature's building blocks

>> **Astronomy + particle physics =** Put away the caffeine because you're already awake: high-energy astrophysics

The field of high-energy astrophysics covers any energetic event in the universe that leads to massive outbursts of energy. If you're wondering what exactly constitutes an extremely energetic event, you're in good company — the definition of "extreme" on Earth completely depends on your own personal assessment of excitement and risk. For example, some might consider walking alone through a park at night to be risky behavior, whereas others need to scale Mount Everest before feeling like they've stepped outside of their comfort zone.

Some events on Earth are less objectively extreme. Torrential rains and heat waves lasting for weeks are examples of extreme weather; skydiving and highlining will have most people reconsidering their life choices. Any event outside of your normal realm of daily experience can be thought of as "extreme."

In the world of astrophysics, events generally considered to be high-energy are subject to more objective considerations. The universe as a whole contains literally billions of stars and other objects. When seen collectively from an energy emission point of view, a few classes of activities rise to the top. The highest-energy-producing events currently known to science include supernovae, emission jets from active galactic nuclei, and gamma-ray bursts. These are objects that meet the high-energy criteria because they produce very high amounts of energy in relatively short periods of time.

Triple E: Extreme energetic events

A typical low energy event in our world might be sitting in your favorite recliner and watching a sports game on television (baseball — that's the one with the hoops, right?) or lazing around on the couch following a big meal. Or both! Low energy events both use and give off very little energy. In contrast, high-energy events emit massive amounts of energy. Here, density matters. Some of these events are related to remarkably small yet dense objects, like merging neutron stars, while others are related to how the entire — again very dense — core of a galaxy shines.

There's a wide range of possible areas of study within the field of high-energy astrophysics. One topic generating lots of current interest is cosmic-ray acceleration, a concept that's exactly what it sounds like — cosmic rays can be accelerated by the aftermath of high-energy events such as supernovae explosions.

TECHNICAL STUFF

Backing up briefly, cosmic rays are a type of ionized particles; they're high-energy fragments of atoms, typically atomic nuclei, coming from outside of our solar system. They're speedy little demons that move fast enough to approach the speed of light (a cool 300 million meters per second, or 186,000 miles per second).

A variant of cosmic rays, ultra-high energy cosmic rays (UHECRs), are even more energized subatomic particles. UHECRs are typically protons and have energies ranging from 10^{15} eV to 10^{20} eV.

REMEMBER

An eV is an electron volt, and it's used to measure the energy of an electron as it moves through a 1-volt potential. Think of it as a way to compare energy units. Cosmic rays can be created outside our solar system and, as Figure 12-1 shows, many are deflected by the Sun's magnetic field. This field extends out into a vast zone surrounding the solar system called the heliosphere; however, a few do make it through to be detected here on Earth.

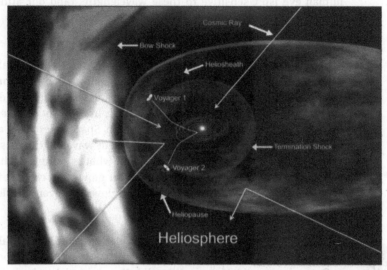

FIGURE 12-1:
Artist's conception of cosmic rays in the heliosphere.

Courtesy of Walt Feimer/NASA GSFC's Conceptual Image Lab / Public Domain

Cosmic rays likely originate from supernovae, pulsars, gamma-ray bursts (more on these later in the chapter), AGN (active galactic nuclei), massive clusters and stars, and other energetic gathering spots. When a supernova explodes and jettisons remnants out into space, the x-ray or gamma-ray emissions from these remnants can speed up or accelerate the energy of nearby cosmic rays. Cosmic rays generated from these galactic events crash into the Earth's atmosphere, breaking up as they interact with other atoms.

How do we know? ALMA, Hubble, JWST, Chandra, and others

Low-energy events are extremely common and can be witnessed with the unaided eye. Look up at the night sky on any night, for example, and you see relatively low-energy stars all around you. Because low-energy events often give off light in the visible wavelength range, they are detectable to the human eye. Astrophysicists can use special cameras to observe even longer-wavelength (and therefore lower energy) radiation such as infrared and radio waves.

TECHNICAL STUFF

High-energy astronomical events, on the other hand, are very rare and (lucky for us!) generally occur at extremely large distances. Seeing these events typically requires advanced measurement tools. Our sky is very cluttered in the colors of light our eyes see (and all low-energy activities emit). High-energy events still give off visible light, but they also shine bright in the x-ray and gamma-ray portions of the spectrum. By looking there, we can spot these rare objects and events over all the visible-light noise. Specialized detectors are necessary because the human eye isn't sensitive to these wavelengths (and our atmosphere — again lucky for us — blocks these colors). Once a detection is made, space-based and Earth-based telescopes start gathering data.

High-energy astrophysics observations often start with an alert. Survey telescopes like NASA's Swift Observatory may spot a flash in gamma rays and then follow up using detectors sensitive to x-ray and visible light to identify what made the flash. With a location in hand (or rather, in data), other telescopes can follow up. Although the Hubble and James Webb Space telescopes are both blind to gamma-ray light, they can observe the longer wavelengths these events also give off, allowing us to understand more about what is going on. These lower-energy emissions are tied to gamma-ray burst after glows that can last for days or weeks. In a few very rare instances, these after glows are even visible in your average backyard telescopes!

TECHNICAL STUFF

Another high-energy mission that provides extraordinarily valuable data is the Chandra X-ray Observatory, launched in 1999 into orbit about 86,500 miles (139,000 km) up into space. Why did Chandra need to be launched so far into orbit? Thank our atmosphere, that glorious buffer responsible for allowing us to breathe air, drink water, and not die from solar radiation, but also one that absorbs x-ray radiation. Chandra has to orbit out far past our atmosphere to be able to study those high energy waves. From its perch in space, Chandra can study x-rays given off by supernovae, black holes, galaxy clusters, and other extremely hot parts of the universe.

Ground-based telescopes provide a similarly valuable method of studying the sources of high-energy radiation. ALMA, the Atacama Large Millimeter/submillimeter Array, is a radio telescope array located at high altitude in Chile. Its 66 antennae are able to record radiation in the millimeter/submillimeter range, and its location is no coincidence; the elevation and very dry climate allow ALMA to capture data that would be lost due to water vapor at lower (and wetter) elevations. Although ALMA observes at a wavelength range that's in between infrared and radio waves, its observations have added details to our views of pulsars, quasars, and other high-energy phenomena This is because the same environments that often release gamma-ray and x-ray radiation also accelerate electrons, and those electrons then give off radio radiation.

Prime examples: Active galactic nuclei, gamma-ray bursts, supernovae

In astronomical terms, high-energy events are ones that give off very high amounts of energy. One key example is the active galactic nucleus, introduced in Chapter 10 as the dense, super-bright center of a galaxy that contains a supermassive black hole that is actively feeding. AGNs emit extremely luminous jets as matter circles and is pulled into the central black hole, and they are known as some of the highest-energy emission sources in a galaxy. They can give off high-energy x-ray and ultraviolet radiation that varies on timescales of hours or days. Quasars and blazars are two kinds of AGN, and Figure 12-2 shows an AGN called Centaurus A in both optical and radio wavelengths — huge jets of high-energy particles are visible in the radio image, and can also be observed at x-ray wavelengths.

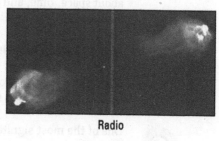

FIGURE 12-2: Active Galactic Nucleus (AGN) Centaurus A, in optical (left) and radio (right).

Optical

Radio

Courtesy of NOAO

Courtesy of NRAO

Another good example of high-energy astrophysics in action is gamma-ray bursts (GRB), or super-high-energy explosions in galaxies far, far away (see Chapter 8 for more info). These emissions in the gamma-ray wavelength range are brief, no more than a few minutes and often much shorter, but they can be as bright as an entire galaxy for that brief instant.

There are essentially two types of GRBS:

>> Short-duration bursts (< 2 seconds)

>> Long-duration bursts (< a few minutes)

GRB are some of the brightest and highest energy events since the Big Bang and are thought to be given off during a supernova or neutron star merger.

REMEMBER

Bringing up the rear in spectacular fashion are supernovae. See Chapter 5 for a general background; in this context, you can see supernovae as a star's ultimate end-of-life experience. Supernovae are massive star explosions, ones that increase the luminosity of a star by thousands or millions of times, and take place when a star has used up its fuel. These explosions can radiate out more light energy than our Sun will give off in its life, and have been recorded in astronomical history over at least the last 2000 years. These brief explosions can produce a neutron star, a black hole, or (most rarely) nothing but an expanding cloud and light. They emit high energy out to x-ray and gamma-ray wavelengths.

Gravitational Lensing

When you hear about the fabric of space-time, do you ever imagine a blanket or quilt with pictures of planets? It's definitely a stretch of the imagination to talk about space, time, and cloth together, but we hope to convince you that the fabric analogy is a good one.

Although we go more in depth in Chapter 15, one way to start wrapping your head around the concept of space-time is to take the dimensions you're familiar with from the known environment — X, Y, and Z — and consider how they're affected by time, the so-called fourth dimension in the fabric.

TIP

One of the most significant theories in the realm of astrophysics, Einstein's theory of general relativity, suggested that gravity is associated with the curvature of space and time. Gravity can actually bend the fabric of space (this is where the cloth analogy comes into play), as shown in Figure 12-3. Imagine this scenario in three dimensions, and you'll have an idea of what the curvature of space-time looks like.

Gravitational lensing is a phenomenon that occurs when the gravity of massive celestial bodies such as galaxies or black holes bends light in such a way that it magnifies and distorts the appearance of other objects. Gravitational lensing, which is a consequence of the bending of the fabric of space, wouldn't be possible without Einstein's theory of general relativity.

The central ideas to understand, in order for gravitational lensing to be possible, are

>> Space and time are not separate entities; they're bound together into space-time.

>> Large celestial bodies such as galaxies can bend, or curve, the fabric of space-time.

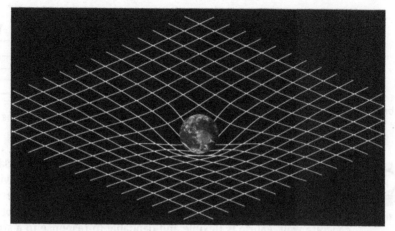

FIGURE 12-3:
Gravity can bend the fabric of space.

Here's a short analogy to visualize how gravitational lensing works. Imagine a scenario where a distant object is behind a galaxy, rendering it invisible under normal circumstances. The gravitational force of the galaxy in the foreground bends light from the distant source toward you — light originally headed toward some other part of the universe — thus enabling the distant object to be viewable. How can this happen? To understand gravitational lensing, let's start with regular lenses.

The bending of light

We can break down gravitational lensing into its component terms: gravity and lens. You might be tired of hearing about gravity by now, but it's pretty difficult to cover most topics in astrophysics without it! Gravity is the force of attraction, in this sense, between two celestial bodies.

TECHNICAL STUFF

A lens is a transparent curved object that's used for either focusing or dispersing rays of light that pass through it. Both surfaces of the lens are polished and curved to either a concave (sinking inward) or convex (bulging outward) shape (see Figure 12-4 for examples of both). Lenses focus incoming light rays via refraction, the bending action which causes the light rays to change direction as they pass through the lens' curved surfaces. You very likely use lenses every day — common applications of individual lenses include contact lenses and eyeglasses, whereas combined or compound lenses are used in optical equipment such as telescopes and cameras.

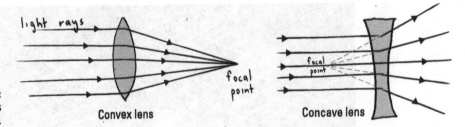

FIGURE 12-4:
Light as it bends
through a lens.

TIP

Light being bent by a glass lens is a typical subject for a physics class, but light can also be bent by other methods — including, you guessed it, gravity! The gravitational bending of light is made possible by the idea that space-time is a fabric. There's a reason why it wasn't named the space-time brick, or the space-time concrete wall. Fabric is flexible and bendable in response to a stimulus, and this flexibility is what makes the bending-like-a-lens concept possible.

Figure 12-5 shows how gravitational lensing works. Start with a source object like the dark circle at the far left, one that's located on the source plane. Rays of light from the source object travel along the two straight lines shown but then pass by the large dark circle object in the middle that's positioned on the lens plane. When the two rays of light pass by the lens object, they're attracted by the lens object's strong gravity and are subsequently bent. Rather than spreading away from each other, the light rays are bent back toward each other and eventually enter the eye of the observer.

Our observer friend, though, doesn't initially know anything about the object in the lens plane and instead assumes that the two light rays had traveled in straight lines. Light rays can be traced all the way back to the source plane as the dotted lines at top and bottom, making it look like there are two identical copies of the source object. These apparent copies are shown as the open circles at the top and bottom of the source plane.

The simplified diagram in Figure 12-6 is a two-dimensional model, but in reality, this process would happen in all three dimensions. Instead of just seeing two copies of the source object, the observer would actually see two smeared-out partial rings above and below the location of the source object.

TIP

The net result of gravitational lensing in astrophysics is that it allows scientists to better understand a galaxy's mass and how much dark matter it contains. The amount of bending caused by an object in the lens plane depends on the mass of that object; the more massive it is, whether a galaxy, cluster, or black hole, the more the rays of light from the source object are bent. They also let us see light from sources so far away that we can only see them thanks to the light-magnifying power of gravitational lenses.

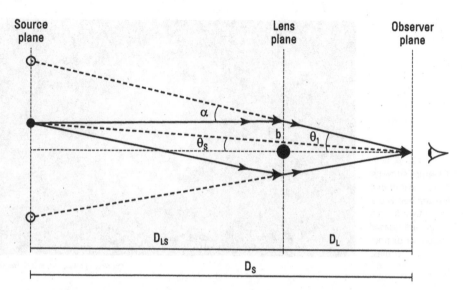

FIGURE 12-5:
Gravitational
lensing and
space-time
curvature.

Gravitational lenses can produce beautiful objects in the sky, such as an Einstein Cross or an Einstein Ring. Figure 12-6 shows an Einstein Cross as seen by the Hubble Space Telescope. A massive foreground galaxy has bent the light from a distant quasar into four separate images. These images appear as the bright spots surrounding the nearer, central galaxy, as shown in the diagram in Figure 12-7.

Gravitational Lens G2237+0305

Courtesy of NASA, ESA, STScI

FIGURE 12-6:
Einstein Cross
G2237 + 0305, as
seen by the
Hubble Space
Telescope.

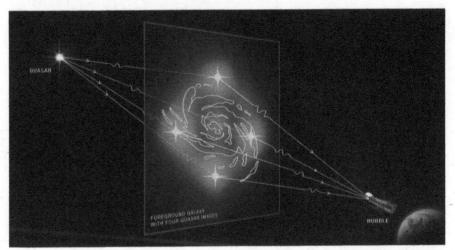

FIGURE 12-7:
Diagram showing how gravitational lensing bends the light from a distant quasar into four distinct copies.

In some cases, the light from a distant object can be spread into long arcs that can form an entire ring. Figure 12-8 shows Einstein Ring SDSS J0146-0929, as seen by the Hubble Space Telescope. This massive galaxy cluster is so big that it's bending space-time around it, causing the light from distant galaxies in the background to be spread into arcs as observed from Earth.

Strong, weak, and microlensing

TIP

Gravitational lensing can produce some beautiful and puzzling images, but it can also be used to figure out the mass of astronomical objects including galaxies and galaxy clusters. Gravitational lensing can even be used to find exoplanets. There are three main types of gravitational lensing: strong lensing, weak lensing, and microlensing:

>> **Strong lensing:** This type of gravitational lensing produces multiple images of a background object. The lensing effect can make the light from the source object brighter than it would normally be because multiple rays of light are bent and focused on Earth. Strong lensing can reveal source objects that are weaker than you would be able to see without the lensing effect. The amount of bending also depends on the mass and mass distribution of the lensing object. The most distant stars and galaxies were all discovered with strong lenses.

>> **Weak lensing:** This effect is less strong and doesn't produce multiple images of a source object. Instead the source is distorted in a way that allows the

lensing mass — for instance invisible dark matter! — to be measured. Weak lensing can also be used to map out billions of faint galaxy clusters by using distortions in the cosmic microwave background radiation to detect them through their weak lensing effects. The result is an independent way to measure dark energy (see more on that in Chapter 15).

>> **Microlensing:** As its name implies, microlensing is a small type of lensing that can be used to detect exoplanets (see Chapter 7). When one star passes in front of another, as seen by Earth, there's a small brightening in the light from the background star due to microlensing. If the closer of the two stars has planets orbiting it, that alters the brightness increase slightly, allowing astronomers to detect them! This effect will be used by the upcoming Nancy Grace Roman Space Telescope to map out hundreds or thousands of exoplanets.

Heading Down the Wormhole

Congratulations! You've finally made it to the true science fiction part of this book. You're likely familiar with wormholes from a trip to the cinema. Wormholes are featured prominently in television and movies, so much so that most common knowledge about wormholes has been handed to us via Hollywood.

And with good reason, because wormholes (tunnels, or "shortcuts," between two points in space and time) are one of the few celestial entities that have been predicted but never seen. Then again, perhaps we just haven't waited long enough. Gravitational waves, for example, were predicted as part of General Relativity, but they weren't observed until detector technology developed enough to measure the tiny effects they cause. Wormholes also arise from several different aspects of General Relativity and, as such, they could exist.

TIP

If you're wondering why wormholes haven't been proven to exist yet, think about the last time you went on a vacation through time. Yep, that's right — you haven't, because despite what television producers would have us believe, time travel does not currently exist. A central idea behind wormholes is that you could slide in one part of the tunnel and out the other. The result is that you'd end up not only in another place in space, but also in another place in time. Wormholes would allow you to connect two points in space, but not just any two points; you could potentially connect two points that could never be connected otherwise because they exist in different places in time, space, or both. Given that you can't toss a wormhole out of your pocket and hop on over to last year, why are we so sure they exist?

FIGURE 12-8: Einstein Ring SDSS J0146-0929, as seen by the Hubble Space Telescope.

Courtesy of ESA/Hubble & NASA; Acknowledgment: Judy Schmidt

A wormhole by any other name . . . Einstein-Rosen bridges

The answer comes from Einstein's 1915 theory of General Relativity. Wormholes can be considered a special case of the solutions of Einstein's equations of General Relativity.

REMEMBER

Black holes were predicted to exist based on a solution to General Relativity, as mentioned in Chapter 8. From that black hole solution, it turns out that there's an extension fitting the equations where you have two black holes connected together by a throat (a narrow link attached to the two black holes).

TECHNICAL STUFF

Einstein collaborated in 1935 with Nathan Rosen, an American physicist, on how general relativity could explain this bridging of time and space. The conceptual outflow of these conversations was dubbed the Einstein-Rosen bridge, but they're much more commonly known today as wormholes. One of the weirdest things about wormholes is that they exist in the full four dimensions of space-time. They act as a shortcut not just between two locations in XYZ space, but also between two locations in time. In other words, a wormhole needs to allow backward (or forward) travel in time. Figure 12-9 shows what the geometry of a wormhole may be like.

Would wormholes be too unstable to last long enough to be detectable? The equations show that they would be very unstable. In order for a substance to keep one open it would need to have a negative energy density, and no currently known material does. Even if this kind of material could be found, however, British

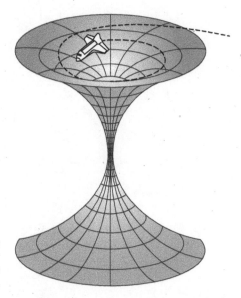

FIGURE 12-9:
The inner
workings of a
wormhole.

astrophysicist Stephen Hawking argued that time travel would never be possible. In his models, if a particle ever actually entered a wormhole, that action would destabilize the wormhole so much that it would quickly collapse. The particle wouldn't have a chance of traveling from one end to the other. And since space is permeated by light and cosmic rays . . . it's hard to imagine a wormhole lasting long enough for us to observe.

Making connections between wormholes and string theory

TIP

If you're a guitar player, or a violinist or cellist, you know that when you pluck or bow a string on your instrument, those strings vibrate and produce sound. Now imagine that rather than creating sound, vibrating a string creates particles. That, in a vastly oversimplified nutshell, is string theory, which is an idea that comes from theoretical particle physics.

String theory might allow for the existence of stable wormholes — if it proves true. This involves the use of cosmic strings, fracture-like remnants from early in the history of the universe that astrophysicists theorize were created when the four fundamental forces of nature (strong, weak, electromagnetic, and gravity) separated from each other. In theory, cosmic strings could be strung through the open end of a wormhole and anchored long enough to keep it open and permit travel through the wormhole. Of course, cosmic strings themselves are a theoretical prediction for which we also have no direct evidence. In this light, wormholes can be seen as possible evidence for string theory, and that's enough to excite many astronomers.

4

Cosmology: The Beginning and the End of Everything

See how it all began, starting from the Big Bang. Learn how fundamental particles and the first elements and molecules formed in the very early universe.

Illuminate the formation of the very first stars, discover stellar populations and the reionization epoch, and watch the first galaxies form.

Dive into general relativity in order to understand why the universe may require both dark matter and dark energy.

Understand different theories for the end of the universe. Grasp the Big Freeze, the Big Rip, and the Big Crunch, and keep an eye on future observations that may help determine the most likely scenario.

Chapter **13**

The Big Bang: How It All Began

I f a tree falls in the forest and there's no one there to hear it, did it make a noise when it fell? Experience, perception, and science all play into this thought experiment, one that has no definitive answer except the one that makes sense given what you know about trees and sound.

Understanding the Big Bang takes a similar path. At its highest level, the Big Bang theory is an explanation for the creation of the universe. The Big Bang begins from a single, explosive point. Emanating from that point is the creation of matter, space, time, and everything else in our known universe. Astrophysicists combine astronomical observation with physics and mathematics to arrive at what most believe is a valid theory as to how our universe began. Is it the only possible explanation? No, and this chapter will cover a few alternatives, but the Big Bang is the proverbial rack where astronomers tend to hang their hats.

REMEMBER

Certain leaps of faith will be required when wrapping your brain around the Big Bang. And no, we're not talking about religion, though it should be noted that the line between theoretical cosmology and the philosophy of religion can be very thin. Not only do we explain the science behind why astronomers believe the Big Bang took place, but we also show how observations and data support this conclusion.

Here we go!

What's the Point? A Primer on Cosmology

Creating something out of nothing. That pretty much says it all.

<poignant pause>

TIP

Okay, so there's a bit more to it. The field of cosmology (not to be confused with cosmetology, or the art of beautifying your hair, skin, fingernails and toenails; we know absolutely nothing here) is the field of study that seeks to understand the origins of the universe. Cosmology is a specialized branch of astrophysics, but both include the study of space, stars, galaxies, and everything outside of the Earth's atmosphere. There's so much crossover between the two fields that it's hard to draw an exact distinction. In this context we'll focus on a specific period of time, starting with the initial Big Bang and spanning out about 500 million years to when the first stars were formed.

Let's get right to it: The Big Bang

TIP

Time to deep-dive into the Big Bang theory, so let's begin at the beginning. But first, what do astronomers think came before the Big Bang? Nothing. Honestly, nothing — the whole idea of the Big Bang is that it is a single moment in time when the universe began. Before the Big Bang, there was nothing at all — or at least nothing that you could sense or study or measure in any way. See Chapter 16 for a few more speculations on this topic.

The initial concept for the Big Bang theory began with Einstein's theory of General Relativity, published in 1915. See Chapter 15 for more on the whole story, but the basic idea is that Einstein initially included a constant in his equations to keep the universe stable and static, in keeping with the views of the time. However, in the 1920s, Russian physicist Alexander Friedman found solutions to Einstein's laws of universal gravitation that predicted that the universe could actually be expanding or contracting.

Soon afterward, Belgian astronomer Georges Lemaitre suggested that in an expanding universe, the starting point of that expansion, roughly 13.8 billion years ago, was one single point that was all of space, as conceptually illustrated in Figure 13-1. This diagram shows how space and time expand out from a single starting point.

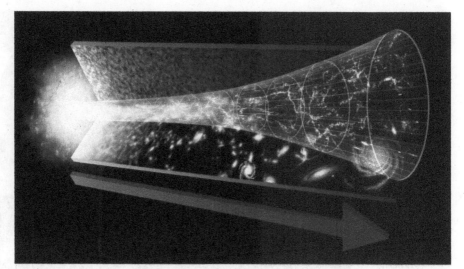

FIGURE 13-1:
Artist's conception of the universe beginning at a single point in the Big Bang. The arrow shows the direction of time.

Lemaitre, who was also ordained as a Catholic priest, argued that this single point, also called a singularity, was dense enough to trigger an explosion. And not just any explosion, but one with so much energy and force that it created matter. Prior to the explosion, there would have been no space; the initial explosion would have created both space and matter. Now, here's the weird thing — there was and is no center to the universe. Astrophysicists now believe it is just an expanding volume called a hyper-toroid that somehow loops back on itself. Put simply, in two dimensions plus time, we live on the surface of an expanding donut, and that surface has no center.

Einstein was originally resistant to the idea of an expanding universe, but when Hubble's observational evidence for an expanding universe, based on his redshift observations of galaxies, was published in 1929, Einstein was finally persuaded. See Chapter 15 for more on the story, including how Einstein considered the inclusion of the Cosmological Constant his greatest mistake.

TIP

Ironically, although Lemaitre is sometimes called the "Father of the Big Bang," the term "Big Bang" didn't come from Lemaitre. It was instead coined in 1949 by an English physicist named Fred Hoyle in an off-the-cuff remark comparing it to his preferred theory of universe creation, the steady state model. More on this topic below.

We'll cover each of the steps immediately after the Big Bang in more detail later in this chapter, but here's a quick overview of the story, which is summarized in Figure 13-2.

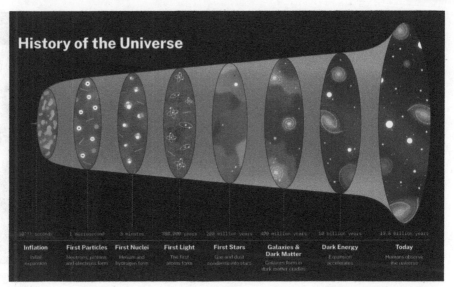

FIGURE 13-2: Timeline of the Big Bang.

Inflation	First Particles	First Nuclei	First Light	First Stars	Galaxies & Dark Matter	Dark Energy	Today
Initial expansion	Neutrons, protons, and electrons form	Helium and hydrogen form	The first atoms form	Gas and dust condense into stars	Galaxies form in dark matter cradles	Expansion accelerates	Humans observe the universe

Courtesy of NASA / Public Domain

Expansion of space, not your grocery bill: Cosmic inflation

As that initial singularity exploded, every point in the new universe would have expanded exponentially in all directions. It also would have expanded very quickly — at speeds up to the speed of light — and this expansion is known as the theory of cosmic inflation. The main idea of this theory is just what it sounds like. Picture the universe like the surface of an empty balloon, but without the balloon. Inflation is when the not-balloon expands everywhere in all directions and creates the spatial extent of the universe in the process. Just before this period of inflation, cosmologists think that the four fundamental forces (gravity, electromagnetism, and the strong and weak nuclear forces) were all briefly united in a blaze of glory. As the forces separated, they gave off a huge burst of energy which initiated cosmic inflation.

Creation of fundamental particles

TECHNICAL STUFF

At the outset of the initial Big Bang explosion the universe was hot — unbelievably hot, millions and billions of times hotter than our Sun. During this period of intense heat, the only particles that could have formed would have been electrons and quarks (see Chapter 2 for a primer on fundamental particles). Soon those infant quarks would have combined into subatomic particles such as neutrons and protons, ones that then combined into fundamental elements such as hydrogen and helium. Within the first three minutes, there were even some nuclear reactions that generated a bit of lithium and beryllium. It would take the universe roughly 400,000 years to cool to about 5400 °F (3000 °C). At this point, atoms would have formed via a process called "recombination," and from this point on it's thought that the Cosmic Dark Ages began.

Too early to shine or twinkle

REMEMBER

Here on Earth, you've likely studied the Dark Ages as the Middle Ages, a period in time between about the 5th and 14th centuries CE. This period is much debated by historians; it represented a slowing-down of development in art, architecture, and cultural innovation between Classical Greece and Rome, and their massive rebirth in the Italian Renaissance. The Cosmic Dark Ages, on the other hand, refers to the period where the universe would have been shrouded in darkness because there was too much neutral hydrogen to enable the light from the earliest stars to be transmitted through the universe.

During this period, matter (hydrogen, helium, a bit of lithium and beryllium, but not quite dust yet) began clustering and forming into the universe's first stars. Stellar formation was fast and furious, galaxies and clusters started coming together, and all manner of celestial objects started coming into play.

If we had x-ray goggles that could somehow look back in time, we might be able to study these objects; they weren't visible until that big cloud of neutral hydrogen lifted. The "great reveal" came about a 300 to 600 million years post–Big Bang in what is called the Cosmic Dawn. We go into all of those primary stellar objects in Chapter 14.

Scientific Evidence: Why Do We Think There Was a Big Bang?

As a scientist or someone who appreciates science, evidence is keenly important to you. Theories and suppositions are great but at the end of the day, you're going to want proof. Hardcore tangible evidence is hard to come by in astronomy because although you can see many celestial bodies from your own backyard, you can't touch or hold them. You can't study them in detail without the aid of Earth- or space-based telescopes and observatories. So why should you believe that the Big Bang ever happened?

TECHNICAL STUFF

First, remember that we talk about how looking at distant galaxies and other objects is also looking back in time because the speed of light is fixed. Our observable universe is the spherical region from which light has had time to reach us during the 13.8-billion-year age of the universe.

REMEMBER

Instruments such as the FIRE IR spectrometer on the Magellan telescope in Chile, a sophisticated device that can make observations at extremely high redshift, have taken observations of the earliest stars and galaxies (see Chapter 9 for a primer on redshift). Using this kind of instrument, you can make observations of galaxies and quasars at 95 percent lookback time to the Big Bang. The implication here is that the universe was only 5 percent of its current 13.8-billion-year age. By combing gravitational lensing, galaxy clusters, and the James Webb Space Telescope, we're starting to look back even further! How can you use observations to determine the existence of the Big Bang? The story continues below.

Hubble's Law and the expansion of the universe

Just a few years after Lemaitre went public with the idea that the universe began at a single point, American astronomer Edwin Hubble used the Hooker Telescope on Mount Wilson in California to study clouds of light in the sky that he called "nebulae." Hubble used redshifts to determine that these objects were moving away from us. He combined these measurements with standard candles to determine their distances, and realized that these distant objects were receding from the Milky Way at a rate proportional to their distance. Not only did Hubble determine that what were previously thought to be nebulae in our own galaxy were in fact distant galaxies in their own right, he also discovered that the farther away a galaxy was from us, the faster it was moving away. This finding was called Hubble's Law, and it laid the groundwork for one of the most amazing discoveries of 20th-century astrophysics: the idea that the universe is expanding, leading back to — you guessed it — the Big Bang theory.

TECHNICAL STUFF

How can the universe be expanding in all directions, and doesn't that mean the Earth must be special if it's at the center? Sorry, but not at all — in fact, if the whole universe is expanding and creating more space around it as it goes, every point is moving away from every other point. Remember that not-balloon that we were inflating, when we talk about cosmic inflation earlier in this chapter? Now imagine drawing some stars on the balloon, and then blowing air into it. As the balloon expands, all the stars on it move away from each other because the surface of the balloon, like the universe, is expanding (but in three dimensions!)

If the whole universe is expanding, then logic dictates turning the clock back and following that expansion back in time. If you go back far enough, the universe would collapse down to a single point. Presto, you just found the Big Bang — the starting point, and event, that put the whole universe into motion and initiated the creation of time and space itself.

Not for popcorn: Cosmic microwave background radiation

A second major piece of evidence supporting the Big Bang lies in the cosmic microwave background radiation (CMB). The CMB was first detected accidentally in the 1960s via what American astronomers Robert Wilson and Amo Penzias thought was background radio interference. It's actually the most distant signal detectable from the universe after the initial energy generated by the Big Bang.

Picture the universe immediately after it was created — it was hot, and we mean really hot! The universe post–Big Bang was also incredibly dense, and most of the energy was in the form of radiation rather than particles of matter. However, because the universe has been expanding ever since the Big Bang, that initial high temperature pulse of energy has been stretched out to longer and longer wavelengths as space itself has been expanding around it. The result is a decrease in the temperature of this energy.

TECHNICAL STUFF

Scientists predicted theoretically in the 1940s that this remnant signal from the Big Bang might be detectable at microwave wavelengths, ones corresponding to a temperature of just a few degrees Kelvin. This signal would also be isotropic, meaning it would be coming from all directions. The prediction makes sense because every point in the early universe gave off light capable of becoming someone's CMB. We see light in all directions, and all that light was sent flying when the universe was about 400,000 years old. If we moved far far away, we'd still see a CMB that would come from all directions — those photons just originated at the same moment from a different set of places.

In 1964, Penzias and Wilson were setting up a new telescope to measure microwave radio signals from the Milky Way. This antenna was one of the most sensitive that had ever been used for this wavelength range. The two astronomers spent months trying to calibrate it to get the best possible signal. Despite their best efforts, they kept finding a background static signal coming from all directions. Finally, after ruling out problems with their equipment and every other possible cause (including pigeon droppings in the antenna!), they came to believe that the signal was real. It's a uniform signal now called the cosmic microwave background, coming from all directions with an effective temperature of about 2.73K.

This discovery is some of the best proof for the Big Bang, and it earned Penzias and Wilson the Nobel Prize in 1978. Figure 13-3 shows the CMB in an all-sky map, as seen by the dedicated Wilkinson Microwave Anisotropy Probe (WMAP) spacecraft over nine years of observations. The European Space Agency's Planck spacecraft has taken even more recent observations of the CMB. Although the CMB is mostly uniform, subtle variations can be seen — more on those later.

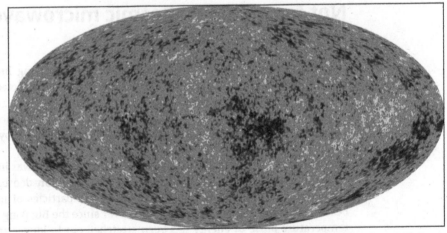

FIGURE 13-3:
The cosmic microwave background radiation, as seen by the WMAP spacecraft.

Courtesy of NASA / WMAP Science Team / Public Domain

The Big Bang Nucleosynthesis (BBN) era

REMEMBER

Another clue in the search for Big Bang evidence is the relative abundance of light elements in the universe. In this case, we're talking about elements like Helium 3, Helium 4, Lithium 7, and Deuterium. You learn in Chapter 2 that elements can have different isotopes, ones that change with the number of neutrons in the nucleus; these are written as different numbers.

TECHNICAL STUFF

According to Big Bang theory, just a second after the Big Bang the universe had a temperature of about ten billion degrees and was awash with fundamental particles and photons By 100 seconds, the universe cooled to a billion degrees and protons and neutrons would have formed and started building the nuclei of various forms of hydrogen and helium. Up until the universe was about 3 minutes old, it would remain hot and dense enough for these early atomic nuclei to experience nuclear reactions. This era is called Big Bang Nucleosynthesis and it's similar to regular nucleosynthesis inside stars (discussed in depth in Chapter 5).

Theories of the Big Bang predict the ratios in amounts of hydrogen, helium, lithium, beryllium, and their various isotopes that would be produced very early in the universe; these amounts are very close to the amounts actually observed, and these observations provide yet another line of evidence supporting the existence of the Big Bang.

Disproving the steady-state model

Although the Big Bang theory is widely accepted as a valid and plausible explanation for the creation of the universe, it's not without its detractors. The main rival theory to the Big Bang is the steady-state theory, or the idea that the universe had always existed and would always exist. According to this theory, any expansion of the universe during its early formative period was always accompanied by newly-created matter, so the net appearance of the universe never changed.

TIP

English astrophysicist Fred Hoyle (the scientist credited with sarcastically inventing the term Big Bang) was an advocate of the steady state model and one of the first to propose it as a creation story for the universe. At the time (in the 1940s), this model seemed much more reasonable than one where the universe began with a crazy explosion.

Advocates of the steady-state model also talked about something they called the Cosmological Principle, an idea stating that the universe was the same at all locations. This idea implied that the laws of physics here, in the Milky Way, were the same as those anywhere else. In fact, they also added another component to make it the Perfect Cosmological Principle, one that not only showed the universe was the same everywhere in space, but also in time.

Cosmology and theology started interacting in the 1960s, with the Pope weighing in to say that the Big Bang theory was more consistent with Christian theology about the creation of the world, and the steady-state theory became associated with atheism. After the detection of the CMB occurred in the 1960s, however, the CMB was interpreted as a direct detection of the remnant radiation of the Big Bang, and the steady-state theory quickly fell out of favor. The Cosmological Principle, however, got to live on in space with a rewording to admit that things change with time.

Making Sense of the Unimaginable with the Theory of Inflation

After the steady-state theory had been disproved, the next step in validating the Big Bang was to tie Big Bang theory into rapidly improving theories from particle physics.

TECHNICAL STUFF

In the 1970s, American physicist Alan Guth suggested that certain physical parameters of the universe could be explained by a period of inflation that occurred right after the Big Bang. This brief but important period only lasted a fraction of a second, but during this time, the universe expanded at a huge rate and gave birth to an array of particle physics interactions. After this period the rate of expansion slowed down, and the inflationary model merged with what was becoming known as the Big Bang theory.

Scientists now believe that this vacuum-decay-based period of cosmic inflation dates to 10^{-32} seconds after the Big Bang. During this brief period of time, the universe grew from being smaller than an atom to a volume that would place our entire visible universe between the Earth and Sun. By the end of the first second, it would be more than a light-year across, and after ten years, our observable universe would have been the size of our galaxy. Cosmic inflation helps explain a few key observations from cosmology, such as

>> Why is the universe uniform in all directions?

>> Why does the CMB have so little variation?

>> Why is the geometry of the universe flat? (Euclidian for the win!)

The flatness problem

If you're riding a bicycle, flatness is not a problem at all; riding a bike on a flat surface is, for most of us, a good thing. In cosmology, however, flatness refers to both the mass density of the universe and the explanation for it approaching critical density.

WARNING

In an expanding universe, mass matters — too much matter and the universe's expansion will rapidly slow down, reverse course, and end up with a Big Crunch due to the inexorable gravitational attraction of all that matter. Too low a mass density, however, and the universe will stretch and expand forever.

We talk about these possibilities for the end of the universe in Chapter 16, but what matters here is a third possibility. The mass density of the universe is very close to what cosmologists call the critical density, an amount right on the balance between contraction and expansion. This scenario creates a situation that's nice for us, because this mass density is what allows stars, planets, and life to exist, but it's problematic for cosmologists because they tend to shun coincidences. Ergo, welcome to what's known as the flatness problem.

REMEMBER

Remember the Cosmological Principle stating that the universe is the same everywhere? Philosophers became involved with these ideas from cosmology and devised the Anthropic Cosmological Principle. Per this idea, the universe must be arranged the way it is because it would be unable to support life in any other way. The mind-bending question of "Why is our universe just right for life to exist?" then becomes, "If the universe couldn't support life, we wouldn't be here in the first place to ask that question." These ideas can lead into either theories of parallel universes (fun for Spiderman and theoretical cosmologists) or intelligent design (a creator, whether alien or religious, tweaked the universe perfectly to allow life as we know it).

Neither of these paths are very satisfying for astrophysicists. Fortunately all around, the theory of cosmic inflation actually helps out. Rather than requiring a massive set of coincidences to get to a density that's close to the critical density, you arrive at exactly this point by looking at the details of the inflationary theory. A short period of inflation, just after the Big Bang, flattens the universe and spreads matter out. Regardless of the starting conditions of the universe, the result is a density that's close to the critical density. In that regard, inflation solves the flatness problem nicely. We talk more about consequences for the ultimate fate of the universe in Chapter 16.

CMB and uniform temperature

Inflation can also help explain another puzzle related to the cosmic microwave background radiation.

WARNING

As if we hadn't already brought up enough problems around the Big Bang, here's a doozy. This one, called horizon distance, requires thinking back to the start of everything. Given that the universe is about 13.8 billion years old, the farthest back that light (and therefore information) could have traveled and spread is a horizon distance of 13.8 billion light-years. However, observations of the CMB show that it's uniform over much larger distances than the horizon distance. Because these distant parts of the universe would have no way to talk to each other over distances longer than the horizon distance, and they can't share information faster than the speed of light, how could the temperature have become uniform over such a vast distance?

Inflation to the rescue! The radiation that was emitted right after the Big Bang has stretched out over time and space to become the CMB. This radiation was emitted into a universe that was much, much smaller than today's universe. Before cosmic inflation took place, the temperature of the CMB would have been equalized over a much smaller volume of space. During the inflationary era's fast period of expansion, that equalized radiation was spread evenly in all directions, stretched, and became the uniform CMB that's observable today. In this regard, observation of (mostly) uniform CMB becomes compelling evidence for inflation itself.

Where galaxy clusters fit in

A final bit of evidence for the inflationary era comes from the distribution of galaxy clusters. The CMB is uniform, but not 100 percent uniform. Scientists think that quantum mechanics produced small but important fluctuations at the very beginning of the universe. It was originally proposed that these initial fluctuations led to small density clusters of primordial matter, eventually becoming formative material for stars, galaxies, and galaxy clusters. However, the fluctuations predicted in the original Big Bang theory weren't sufficient to explain the observed distribution of mass, particularly the size, shape, and distribution of the largest-scale structures like galaxy clusters. (Flip to Chapter 11 for more details on galaxy clusters).

TECHNICAL
STUFF

Again, inflation to the rescue! The process of inflation can stretch out those tiny quantum fluctuations in initial matter distribution to produce much larger space-time wrinkles. It's thought that these wrinkles could have helped set the stage for large structures like galaxies and galaxy clusters. This cosmic wrinkling was predicted theoretically by American astrophysicists George Smoot and John Mather in the 1980s. The prediction was borne out by observations taken first by the COBE satellite, then also in more detail by WMAP (refer to Figure 13-3).

Radiation Dominance in the Radiation Era

After the explosive beginning of the universe with the Big Bang, what happened next? The stars and galaxies did not form until a few million years later, but the universe was not sitting idly by. From subatomic particles to the creation of the first element, these were busy days indeed.

Radiation was king in the years following the Big Bang. Why? Because that's all there was! The major evolutionary areas of the universe have been broken into three major eras, with each then subdivided into smaller chunks of time called epochs. Table 13-1 shows the Radiation Era, the Matter Era, and the Dark Energy Era. We talk about all of these in the sections that follow, starting with the Radiation Era.

TABLE 13-1

An overview of the Radiation Era, Matter Era, and Dark Energy Era

Era	Time Span After the Big Bang
Radiation Era	0 seconds to 50,000 years after the Big Bang
Matter Era	50,000 years after the Big Bang through today
Dark Energy Era	9.8 billion years through the end of time

RADIATION ERA

The Radiation Era has a number of epochs, or specific periods in time and history. Universal evolution moved at a quick pace and the time period for each may seem ridiculously short, but these distinctions are necessary because the laws of physics themselves were changing during this time. Each epoch is described below by key characteristics including the time span in which they occurred, and the temperature and density of the universe at the time. During all of these epochs, the universe was cooling and expanding.

- **Epoch:** Planck Epoch
- **How long after the Big Bang:** From 0 to 10^{-43} seconds following the Big Bang
- **Temperature:** Infinitely hot
- **Density:** Infinitely dense

What happened? The start of time as we know it! A gravitational singularity leads to major forces in the universe (gravity, electromagnetism, strong and weak nuclear forces) either combining into a "fundamental force" or being equal. The laws of physics could have been very different during this time. The universe measures only one Planck length, 1.6×10^{-43} meters, across.

- **Epoch:** Grand Unification Epoch
- **How long after the Big Bang:** From 10^{-43} seconds to 10^{-36} seconds:
- **Temperature:** ~10^{32} K
- **Density:** 10^{95} kg/m³

What happened? Gravity breaks out on its own, though the electromagnetic and nuclear forces remain combined, and the first elementary particles appear. Grand

(continued)

(continued)

Unified Theories (GUTs) such as string theory have been proposed to attempt to explain how electromagnetic and nuclear forces work as one during this era, but without complete success yet.

- **Epoch**: Inflationary Epoch
- **How long after the Big Bang**: From 10^{-36} seconds to 10^{-32} seconds
- **Temperature**: ~10^{30} K
- **Density**: 10^{90} kg/m^3

What happened? The nuclear strong force separates from the electroweak force, giving off a burst of energy that triggers cosmic inflation (the very fast expansion of the universe). The universe very quickly grows from the size of a nucleus to the size of a solar system, and early subatomic particles disperse.

- **Epoch**: Electroweak Epoch
- **How long after the Big Bang**: From 10^{-32} seconds to 10^{-12} seconds
- **Temperature**: ~10^{28} K
- **Density**: 10^{85} kg/m^3

What happened? The initial period of cosmic inflation ends. The electromagnetic and weak nuclear forces are still combined, and the universe is composed of photons and pure energy.

- **Epoch**: Quark Epoch
- **How long after the Big Bang**: From 10^{-12} seconds to 10^{-4} seconds
- **Temperature**: ~10^{27} K
- **Density**: 10^{75} kg/m^3

What happened? The weak nuclear force separates from the electromagnetic force, so the four fundamental forces (gravity, electromagnetic, strong and weak nuclear forces) are now separate. Elementary particles start interacting to create bosons and Higgs bosons, allowing the formation of electrons, quarks, and neutrinos. The universe begins a massive cooling period.

- **Epoch**: Hadron Epoch
- **How long after the Big Bang**: From 10^{-6} seconds to 1 second

- **Temperature:** ~10^{20} K
- **Density:** 10^{50} kg/m³

What happened? The universe continues to cool, allowing quarks to combine and start forming hadrons, subatomic particles made up of gluons, quarks, and anti-quarks. Baryogenesis begins — baryons, a category of hadron, are made up of three quarks, and include particles such as protons which will eventually create matter as we know it. The newly formed protons collide with electrons at high speeds, producing neutrons and neutrinos. Matter and antimatter particles (technically, hadrons and anti-hadrons) are created at almost the same rate and quickly annihilate each other. More matter particles are created and leave a small surplus of hadrons.

- **Epoch:** Lepton Epoch
- **How long after the Big Bang:** From 1 second to 10^2 seconds
- **Temperature:** ~10^{12} K
- **Density:** 10^{16} kg/m³

What happened? Following the end of the Hadron Epoch, some leptons and antileptons survive the hadron / anti-hadron collisions and make up most of the mass of the universe, as well as the surplus of hadrons. The universe consists mostly of neutrinos, electron-positron pairs, and photons that are released as energy. By 10 seconds after the Big Bang, the temperature falls enough that most of these electron-positron pairs are destroyed, leaving a small surplus of electrons. Neutrino decoupling also occurs by the end of this epoch, and neutrinos no longer interact directly with other forms of matter.

- **Epoch:** Nuclear Epoch
- **How long after the Big Bang:** from 10^2 seconds to 4×10^5 years
- **Temperature:** ~10^9 K–3,000 K
- **Density:** 10^4 kg/m³

What happened? The Nuclear epoch is the era of Big Bang Nucleosynthesis. Elements are starting to be created. Protons and neutrons undergo nuclear fusion to become nuclei, and the basic elements of helium and lithium are formed. The temperature also continues to fall, an important detail because the universe began as a ball of hot plasma from which photons can't escape. By the end of this era the universe had cooled to 3000K, allowing hydrogen and helium nuclei to capture free electrons, and photons are able to escape in a pulse of radiation now detectable as the cosmic microwave background (CMB).

The creation of the cosmic microwave background, about 400,000 years after the Big Bang, represents some of the oldest light in the universe that we're able to detect. The theory of everything that happened before is just that, a theory — albeit a good one that seems to explain the observed universe starting with the CMB. By the end of the Radiation era, about 75 percent of baryonic matter was hydrogen, 25 percent helium, with trace amounts of lithium and beryllium. By the end of this final epoch, the fundamental distribution of matter was created. These elements later became the building blocks for the first stars.

Nothing Matters More Than Matter in the Matter Era

The Matter Era followed these previous epochs where the universe was flooded with radiation. Try not to think of these era distinctions as hard and fixed boundaries; the universe did not suddenly go from no matter to nothing but matter. During the years after the Big Bang, matter formed as slow and weak particles. Over the years, these particles gained mass and strength to the point where matter became the prevailing force in the universe instead of radiation, and this is where the transition line to the Matter Era falls. The crossover between radiation domination and matter domination is thought to have taken place somewhere between 50,000 and 500,000 years after the Big Bang. Scientists typically use 50,000 because that's when atoms began to form. And yes, these eras overlap a bit — the formation of the universe was a bit messy!

MATTER ERA

After atoms formed, the heady early days immediately following the Big Bang slowed down a bit as the universe transitioned into a more familiar realm. First hydrogen, then stars, then galaxies started to form and evolve over timescales that can be measured in tens of thousands of years, rather than in fractions of a second. Finally, supernovae exploded and seeded later populations of stars, galaxies clustered together, and a mere 13.8 billion years later, here we are!

- **Epoch:** Atomic Epoch
- **How long after the Big Bang:** From 5×10^4 years to 2×10^8 years

- **Temperature:** 16,000 K–60 K
- **Density:** 6×10^{-16} kg/m^3 and dropping

What happened? The preponderance of matter in the universe exceeds radiation as we tip over into the matter era. Atom formation continues, and electrons attach to nuclei to form hydrogen because temperatures are now cool enough to prevent ionization and loss of electrons.

- **Epoch:** Galactic Epoch
- **How long after the Big Bang:** From 2×10^8 years to 3×10^9 years
- **Temperature:** 60 K–10 K
- **Density:** 10^{-22} kg/m^3

What happened? During this epoch, finally, star formation begins. Quasars develop and initial galaxy creation starts to take place. The universe begins to take a shape that more resembles what you can see today.

- **Epoch:** Stellar Epoch
- **How long after the Big Bang:** From 3×10^9 years to today
- **Temperature:** 10 K–3 K
- **Density:** 2×10^{-25} kg/m^3

What happened? During this period, star formation continues and reaches a high point. Stellar explosions create supernovae and other new celestial objects, and individual galaxies begin clustering into larger, gravitationally bound structures.

TIP

The Stellar Epoch of star formation continues up to this day, and we talk about that in Chapter 14. It overlaps with the next category in the breakdown here, the Dark Energy era, expanded upon in Chapter 15.

Figure 13-4 shows a conceptual diagram of the history of the universe starting with the Big Bang. Over the course of about 13.8 billion years, see how the universe starts with a bright singularity that expands out through inflation and quantum fluctuations. Follow along to the CMB afterglow light pattern, shown at about 380,000 years in this diagram. Next in the diagram comes the cosmic dark ages leading up to the first visible stars at about 400 million years after the Big Bang, and then the development of galaxies and galaxy clusters.

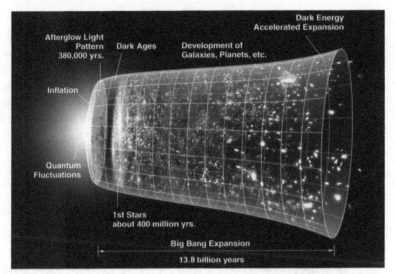

FIGURE 13-4:
Universal
expansion
following the
Big Bang.

Courtesy of NASA / JPL-Caltech / Public Domain

Metric Expansion of Space: The Cosmological Principle

TIP

Now that you know the details of the Big Bang theory, we can wax philosophical. Remember the familiar childhood expression about "assuming" — it makes a donkey's hindquarters out of everyone involved. When deciphering the Big Bang, however, assumptions are absolutely critical. It's equally critical to make good assumptions, ones grounded in science and reason. The Cosmological Principle, as we talk about earlier in this chapter, is a foundational assumption of Big Bang theory, and it comes with two assumptions of its own:

>> The universe is isotropic.

>> The universe is homogeneous.

TECHNICAL STUFF

Anything that is isotropic has the same measured properties in all directions. Homogenous means alike, or similar. As applied to the Big Bang, the assumption is that the universe is homogeneous in the sense that it looks the same in every direction. Taken together, the Cosmological Principle suggests that the universe not only looks the same in all directions, but has approximately the same distribution of matter in all directions. Although the discovery of the CMB disproved the steady-state theory of the universe, the Cosmological Principle still holds steady.

TECHNICAL STUFF

Another relevant concept here is the metric expansion of space. The expansion of the universe actually creates the space that the universe is expanding into, as opposed to the idea that the universe expands and fills up previously empty space. Metric expansion implies that this additional space didn't exist at all until the universe expanded to create and fill it. Let's probe a bit deeper and think about what the expansion of the universe means, and how you can measure it.

We're really not all that special: The Copernican Principle

Copernicus, the famed Polish 16th-century astronomer and mathematician, changed the face of astrophysics forever with the heliocentric theory.

TECHNICAL STUFF

Contrary to previous models which had the Earth at the literal and figurative center of the universe, heliocentrism proposed the opposite: The Earth orbits the Sun instead of the other way around. As Galileo went on to prove, this new principle changed the way humans thought about the universe and our position in it.

As applied to the field of cosmology, the Copernican Principle takes humans down a few pegs by suggesting that there's no one privileged or preferred place in the universe. Observations from Earth can be considered on par with observations from other planets, or even other galaxies, and this idea translates to the concept of an infinitely expanding universe with neither a starting point nor a finish line.

Multi-directional spreading with Hubble universal expansion

REMEMBER

What evidence supports this idea of a continually expanding universe? First, remember Hubble's galaxy measurements. Hubble's Law is the idea that the universe is expanding and that the farther away an object like a galaxy is from us, or the higher its redshift, the faster it's moving away from us. The rate of this continual expansion is known as the Hubble Constant H_o, and it was a critical factor in working backward to determine the age of the universe.

TECHNICAL STUFF

Because the Hubble Constant sounds like it should be — well, constant – does anyone know what it is? Not exactly, but astronomers have been working to measure it. A recently completed 30-year study used the Hubble Space Telescope to measure Cepheid variables and Type Ia supernovae (flip to Chapter 5) as milepost markers to understand the distances to galaxies. This method was used to determine the best estimate yet of the Hubble constant, as $H_o = 73 \pm 1$ kilometer per second per megaparsec. Seems like a great measurement, but there are two problems with it: The expansion rate in the past was likely different from today, and estimates of the Hubble Constant using the cosmic microwave background give a different number.

For the first of these potential issues, measurements with the Hubble Space Telescope of very distant supernovae showed that long ago, the universe was expanding more slowly than it is today. This realization seems contrary to what you'd think: It would seem to make sense that the universe would start out expanding quickly, but that the pull of gravity would start to slow down this expansion rate over billions of years. This line of reasoning would mean that the expansion of the universe should be less today than it was early in history, not more, so the Hubble observations don't seem to make sense.

TIP

So is it possible that the expansion rate of the universe was slower in the past than it is today? Perhaps, and this idea has incited some wild explanations. The simplest way of describing the effect is by adding a new idea to our understanding of the universe. Welcome to the story, Dark Energy! We go into detail on this topic in Chapter 15; the underlying assumption is that dark energy affects universal expansion, causing the expansion to accelerate, and that it comprises about 68 percent of the known universe.

Big Bang radiation and temperature fluctuation

In the second challenge to the estimate of the Hubble Constant using galactic mile markers, can you use temperature fluctuations to tell about the expansion of the universe since the Big Bang? You bet you can! These measurements are another independent way of trying to measure the Hubble Constant by relating it to the cosmic microwave background.

TIP

In the microseconds following the Big Bang, the universe heated to 10^{32} K — compare that to your cozy fireplace! The subsequent cooldown, however, was rapid as all of that initial energy from the Big Bang dissipated into the expanding universe. How do scientists know that these temperature fluctuations during the radiation dominance era actually took place?

>> First, measurements were taken of the temperature of a very distant intergalactic cloud that was found to be in thermal equilibrium with the cosmic microwave background.

- This showed that the radiation from the Big Bang must have been warmer at earlier times.

- There must have been uniform cooling of the CMB over billions of years, which is evidence for the metric expansion of space.

>> Second, the standard cosmological model can be used to measure the value of the Hubble Constant at the time the CMB was given off — remember, that's about 400,000 or so years after the Big Bang.

- This model is called the Lambda Cold Dark Matter theory (ΛCDM – more on this in Chapter 15), and includes detailed models of the CMB and its small-scale variations in temperature and polarization

- Astrophysicists used high precision measurements of the CMB from the Planck mission, and derived a value for the Hubble Constant of about 67.5 ± 0.5 kilometers per second per megaparsec.

Constant expansion, but of an inconsistent rate

So what's going on? Astronomers used two highly precise methods to determine the Hubble Constant H_0, and got two different values:

>> Studying galaxy distance and velocity: $H_0 = 73\pm1$ km/sec/megaparsec

>> Measuring cosmic microwave background radiation (CMB): $H_0 = 67.5\pm0.5$ km/sec/megaparsec

TIP

These methods gave two different results for H_0, 73 vs 67.5 km/sec /megaparsec, and the uncertainties aren't large enough to include both values. Would you expect that both methods arrive at the same pace for universal expansion? Perhaps, but there can be discrepancies between results from these methods, and those discrepancies raise new questions about the fundamental nature of the universe. Just as the galaxy-based measurement of the Hubble Constant ended up showing that the expansion of the universe was not slowing down as expected, perhaps this new H_0 discrepancy could also lead to some new fundamental physics.

In starting to look into this discrepancy, astronomers have investigated recent redshift results and realized that measuring galaxy velocities using redshift might not be as straightforward as has previously been thought. Redshifts have been used for decades to measure the velocity of galaxies both directly and indirectly by observing supernovae in those galaxies. Supernova explosions have an additional component of velocity, however, which had been modeled; new research found that the galaxy and supernova results actually gave different values of the Hubble Constant if they were separated out rather than mixed together into overall measurements of redshift. The significance? It's possible that the redshift method requires a better model of supernova velocities than used previously – and this could mean that one of the estimates for the Hubble Constant also needs to be revised.

The future of direct measurement

Does the future of scientific innovation look promising for more directly being able to measure universal expansion? Absolutely — and in this case, bigger is almost always better. The SKAO, or Square Kilometre Array Observatory, will be a radio telescope array in South Africa and Australia. When fully constructed and operational, this array will be sensitive to radio light at higher resolution and in more wavelengths than any similar facility today. Among the planned projects is a key study of dark energy and cosmology; this study may be able to retrieve data from the neutral hydrogen formed in the Matter Era following the Big Bang. Such data could provide not only the answers to questions about universal expansion, but may also answer many other open questions about cosmology, galaxy evolution, and more.

TIP

Another up-and-coming observatory that could provide mind-blowing answers to basic cosmology questions is the aptly-named ELT, or the Extremely Large Telescope, in Chile. Just as SKAO will be the largest radio array to date, the ELT will be the largest near-infrared/optical telescope. Scientific goals of the ELT include studying the very early cosmos, dark matter, and star formation.

Future observations by the James Webb Space Telescope will also shed new light on the history of the universe, allowing us to see further back in time to earlier in the universe than ever before. It may also allow us to observe some of the first stars in the universe — more on those in Chapter 14!

Chapter **14**

First Light in the Universe, or How a Star is Born

1 3.8 billion years ago, in a galaxy far, far away . . . Whoops, there were no galaxies then. Let's try that again.

13.8 billion years ago, when dinosaurs roamed the Earth . . . Whoops again. Earth is only about 4.5 billion years old, and dinosaurs didn't come along until the Mesozoic era (about 65 million years ago).

Third time's the charm: 13.8 billion years ago, the Big Bang was the trigger that exploded our universe into being. It kicked off a chain of events that ultimately led to the universe's expansion and population with matter and celestial bodies. Chapter 5 sets the stage for how stars are created in today's universe; Chapter 13 covers the epochs and eras that defined the universe's initial expansion, heating, cooling, and matter formation. There's a bit more to the story, though.

REMEMBER

The conditions and environment that facilitate star formation today didn't exist right after the Big Bang, so a different set of rules had to apply to those very first stars. This chapter explores the conditions that let the Patient Zero of star formation come to light (literally!), as well as the galaxies and other structures that followed.

The Cosmic Dark Age

Not so fast, says our stellar Numero Uno. Stars didn't form right at the onset of the Big Bang because there was no matter for them to form from. Also, it was hot right after the Big Bang. Really hot, as in 10^{32} K or about 1.8×10^{32} °F. The universe cooled exponentially over the first few seconds, then more over the subsequent thousands and millions of years. This cooling allowed quarks to form and, eventually, the generation of other subatomic particles. Around 380,000 years later, conditions were right for electrons to bind into atoms — great start! — but not enough for star formation.

The Big Bang cooled the heat

REMEMBER

The initial cooling of the universe after the Big Bang began almost as soon as the initial explosion (see Chapter 13). Cooling allowed nature's building block, the quark, to form. Electrons came next, followed by protons and neutrons. After neutrons and protons formed into nuclei, the environment was ready for the eventual creation of atoms.

During the Radiation era and into the Matter era, hydrogen was the prevailing form of matter. There was too much energy for electrons to stick to these nuclei to form atoms, but not enough energy for the nuclei to fuse together to form heavier elements. So much hydrogen effectively made it impossible for light from early stars to escape. This period in the development of the cosmos is called the Cosmic Dark Ages because there was no way for light emitted from stars at the time to make it through that fog or be detectable with any Earth- or space-based telescopes.

TIP

As the universe kept expanding and creating more space around it, the initially super-hot dense material became more and more attenuated, and eventually began to cool. The first atoms needed a temperature below a few thousand degrees to form and be stable. Because the universe started out at a temperature of hundreds of millions of degrees, it had a long way to go!

Formation of neutral hydrogen atoms and cosmic background radiation

When protons and neutrons were available and ready to become friends in an atomic nucleus, there was a good amount of wait time — roughly 380,000 years — before electrons were able to be captured into orbit around nuclei; from this point, atoms were born.

REMEMBER

Recall from Chapter 2 that hydrogen atoms are combinations of subatomic particles structured around a central nucleus composed of a proton (positively charged) and sometimes a neutron (no charge) or two, surrounded by at least one electron (negatively charged).

Around 380,000 years after the Big Bang, the photons that had been swirling through the cosmos were starting to slow down. Radiation density (in the form of these photons) dropped below the matter density as the universe expanded enough to stretch out and lower the energy of the photons. When these photons no longer had sufficient energy to split hydrogen, the universe underwent a transition from protons and electrons flying freely, to becoming neutral hydrogen (see Figure 14-1). The end result? A simple atom with one proton and sometimes one or two neutrons in the nucleus, and one electron in orbit. (And also some similarly neutral helium atoms and traces of lithium and beryllium.)

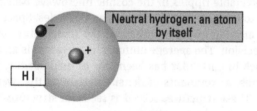

FIGURE 14-1:
A neutral hydrogen atom.

Neutral hydrogen: an atom by itself

HI

And what happened to those primordial photons that used to fill up all available space in the universe? With the creation of neutral hydrogen, many of those photons were finally free. For the first 380,000 years after the Big Bang, all the free electrons kept interacting with the photons; that interaction kept them from traveling very far. When the temperature dropped enough for the electrons to bind together with nuclei to form atoms, the electron soup was no longer distracting the photons, and they could continue on their merry way out into the universe. These photons are detectable in the form of the cosmic microwave background radiation (CMB) — see Chapter 13 for more on the importance of this signal.

HYDROGEN VERSUS NEUTRAL HYDROGEN

Are neutral hydrogen atoms and "normal" hydrogen gas the same? No, they have different structures. The neutral hydrogen atoms created in the aftermath of the Big Bang are composed of a single proton and electron, dubbed H I in atomic terms. The molecular hydrogen you're more familiar with is two hydrogen atoms bound together, or H_2. And don't get confused — there are also H II regions, which are nebulae made of ionized hydrogen. Ionized hydrogen has lost its electron, so there's only a nucleus consisting of a single proton (and sometimes that interrupting neutron . . . or two). It has a positive charge, so is sometimes written as H^+.

A ripple in the universe led to galaxies

The cosmic microwave background radiation initially created by our photon friends is still with us today. It travels the universe and arrives on Earth from all directions and is the oldest visible source of electromagnetic radiation that's possible to detect (see Chapter 13). Although observation has shown it to be relatively consistent in its brightness, it does contain irregularities (called *ripples*) that provide astronomers with hints about how the cosmos evolved in the aftermath of the Big Bang.

TECHNICAL STUFF

In many cases, detectable ripples in the cosmic microwave background radiation can be precursors to the universe's first stars and galaxies. Space telescopes such as COBE, WMAP, and Planck have measured the temperature of the CMB with unprecedented precision. The average uniform temperature is 2.726 Kelvin (about −270°C), but Planck in particular has been able to map out tiny variations, only fractions of a degree, as remnants of density variations that were present right after the Big Bang. These structures served as seeds for structures large and small, from stars to galaxies to galaxy clusters.

End of an era: The cosmic dawn

The Dark Ages that preceded the Renaissance (here on Earth, just to be clear) lasted for about 900 years, from the 5th to the 14th century CE. It took the cosmos hundreds of millions of years longer to allow the light of early stars and galaxies to shine through.

The cosmic dark ages are characterized by elementary particles created in the Big Bang, ones that came together to create neutral hydrogen. At that time there was no opportunity for light to find its way through those particles, but initially there weren't any stars to give off light! As gravity took hold and allowed the very first stars to form, ultraviolet light was able to start breaking up the hydrogen gas and making space transparent again.

This period, lasting from about 50 million to 600 million years after the Big Bang, brought the first stars, galaxies, and even black holes into formation, and this period is called the *Cosmic Dawn*.

Just as the cosmic microwave background is a detectable signal that dates to the formation of the first atoms, it's possible to see the Cosmic Dawn shining in James Webb Space Telescope images. Future radio telescopes may give us further insights by observing the spin-flip radiation given off by hydrogen during this early period of the universe's history.

But first, we're finally ready for some light — so how did those first stars form? Let's see!

Early Star Formation

After those clouds of neutral hydrogen began to cool enough to allow them to collapse under gravity, it was time for the earliest stars in the universe to shine. Chapter 5 covers the basics of star formation; dust and gas, who should have a seat at your Thanksgiving table by now, gather and compress with the help of gravity. The result, after a mere million years or so, is a young star ready to get on with its life.

What about the very first stars in the universe? How were they created? And how were they different from today's stars?

Origins in primordial gas

We start by taking a look at their environment. Stars today (and by *today* we mean any time in the past 10 billion years or so) form in molecular clouds, ultra-dense parts of the interstellar medium, and in protostellar nebulae. These star formation regions are rich in the necessary elements:

>> Hydrogen: ~70%.

>> Helium: ~27%.

>> Heavy metals: ~3%.

Remember that astronomically speaking, *metal* is anything heavier than helium. Heavy elements wouldn't exist until the first stars had lived and died; only hydrogen and helium existed in abundance. Conditions for star formation at the time did not match the conditions in our universe today, and it became clear that a different set of rules applied for the formation of these very first stars.

Following the Big Bang, fluctuations in the density of the primordial gas collapsed due to gravitational attractions and began clumping together. These regions gradually cooled and collapsed enough to become so high-density that nuclear fusion began in cores, more on that later — and Hello World! The first stars were born.

Nuclear fusion to the rescue

TECHNICAL STUFF

As part of the emergence process from the cosmic dark ages, denser areas in the primordial gas began to feel the effects of gravity and compression. Compression led to collapse, and this collapse led to clumping of that neutral gas. A byproduct of these areas' continued compression and heat loss was nuclear fusion (see Chapter 5 for more on how this process works).

Computer models that simulate the formation of these first stars suggest that as temperatures post–Big Bang cooled, hydrogen atoms were able to form molecular hydrogen. This hydrogen emitted infrared radiation as it collided with other unbound hydrogen atoms. The radiation then cooled the local parts of the gas cloud, resulting in a decrease in the gas pressure in these regions that subsequently allowed gravitational attraction to clump them together.

One of the big differences in this period of star formation versus those that happen today is the role of dark matter. Dark matter halos — spheres of dark matter that are densest in their centers — formed first, and regular matter was gravitationally gathered up to form those first star-forming clouds of material. Without the dark matter clumps, there might not have been the necessary pull to start matter collapsing when it did. Today, our galaxy — as every galaxy — sits in an invisible halo of dark matter that extends far beyond our galaxy's visible stars.

Star Classification: Population III

Astrophysicists use sorting and classification to help identify the night sky, and one of the main systems of classification for stars is their Population. A star's *population* refers to its *generation*, a term that correlates closely to its material composition.

TIP

What do generation and composition have to do with each other? Plenty, it turns out, because the composition of star-forming nebulae changed considerably as a result of the formation (and ultimate demise) of the earliest stars. Note that scientists use an inverse ordering system for these populations, so Population III stars are older than Population I. Table 14-1 compares and contrasts the attributes of these different stellar classes in the vicinity of the Milky Way.

TABLE 14-1: **Characteristics of Population III, II, and I Stars in the Local Group**

Population	Relative age	Metal Content	Relative Size	Location
Population III stars	Oldest	Essentially no metal content	Larger than Population I; 100 to 1000 times more massive than the Sun	Not present in modern universe.
Population II stars	Middle-aged, though still called "Old Stars," formed in last 1–15 billion years	Relatively low metal content	Only smaller; 0.8–1.0 times the mass of the Sun; exist in our galaxy	In the Local Group; spiral galaxy halos, elliptical galaxies, globular clusters; primarily in stellar halo and bulge.
Population I stars	Youngest, formed in the last 1 million to 10 billion years	High metal content	About 0.1–10 times the mass of the Sun, with some up to 100	In or near spiral galaxies such as the Milky Way; primarily in disks and spiral arms

No room for diets: Very early stars = massive, low metal content

The earliest stars formed after the Big Bang are classified as Population III stars. Because the universe consisted entirely of primordial hydrogen and helium (with bits of heavier gasses such as lithium), these early stars were the first to use these gasses.

TIP

Although there are distinct advantages in life to being an early adopter of technology, in this case the early stars couldn't reuse heavy metals spat out by the formation of stars that came before — they *were* the stars that came before! As a result, Population III stars had very low metal content and little means of cooling their cores, so they tended to be extremely massive. How massive? Up to hundreds or thousands of times more massive than our Sun!

TECHNICAL STUFF

This large size for low-metal Population III stars is in contrast with today's universe, where dust grains and clumps of material that contain heavy elements can cool star-forming gas clouds down to a temperature of about 10 Kelvin (−441°F). In order to collapse under its own gravity, a clump of gas in a cloud must have a mass called the Jeans mass.

REMEMBER

In Chapter 5, we show that the Jeans mass is directly dependent on the temperature of the gas, and inversely dependent on the pressure of the gas. Population III star-forming regions had gas pressures similar to today's star forming regions. Due to the relative lack of metals, the temperatures of the gas clouds were closer to 300 Kelvin (80°F). The Jeans mass in this case is about 1,000 times larger than for today's stars. Because of this, Population III stars likely started out much larger than the Sun, at least 50 to 200 times larger, and stayed that way throughout their lifetimes.

Because they were so massive, Population III stars lived relatively short lives. It's thought that they burned through their fuel supplies quickly and may have lived only several million years. They would also have been very hot, and most of their sunlight would have been in the form of ultraviolet radiation.

TIP

Can you observe a Population III star? To date, none have been observed directly. Population III stars are theoretical because they existed so early in the universe and had such short lifetimes, so it's believed that none currently still exist. However, some astronomers believe that some of the subtle variations visible in the cosmic microwave background might be from the ionizing light given off by supernovae from early Population III stars. Any future detections of this stellar population are most likely to be through this or other indirect methods, such as gravitational lensing.

Supernova explosions created heavier elements

REMEMBER

All good things must come to an end, and so went the story of the first Population III stars. A star's end of life can have several possible scenarios, ranging from white dwarf all the way up to explosive supernova. Many of the massive Population III stars are thought to have undergone nucleosynthesis and exploded via supernova or black hole collapse. Not so unusual with stars today, but billions of years ago these initial supernova explosions would have been incredibly significant because those explosions created the heavier metals necessary for standard star formation. The supernovae bursts of Population III stars effectively created the first elements heavier than hydrogen and helium that the universe had ever seen.

REMEMBER

Note that not all Population III stars ended up as supernovae. Any stars greater than 250 times the mass of our Sun at the end of their lives would likely have collapsed into black holes. Because these were the early days of the universe, it's likely that these black holes merged with other black holes over time, perhaps becoming concentrated in the central portions of newly forming galaxies and helping seed the formation of the supermassive black holes that exist in most of today's galaxies. The radiation given off by material falling into these black holes could also have led to mini-quasars, resulting in an extra source of light and radiation early in the universe.

More metal: Next-gen stars had carbon, oxygen, iron, and heavier elements

Ever heard the expression, "you are what you eat"? We don't mean that in a literal sense when it comes to the human diet, fortunately, but it does ring true for early star formation. Population III stars were composed nearly entirely of hydrogen and helium because those were the materials available in the post–Big Bang environment.

TECHNICAL STUFF

When Population III stars exploded into supernovae, they started off with just hydrogen and helium. Their supernova explosions were only able to create elements like carbon, oxygen, neon, magnesium, silicon, and iron — because heavier elements generally form either from neutron star mergers or from the end stages of longer-lived, smaller stars such as white dwarfs, neither of which were around during the Population III days.

Although this initial set of elements definitely spiced up a universe that was almost all hydrogen and helium, it certainly doesn't represent the whole spectrum of materials created in today's modern universe. Some of these materials had to be incorporated into the second-generation Population II stars (with the additional requirement that they went the way of the supernova) in order to continue the chain of nucleosynthesis.

As the metal from these first stars was dispersed throughout the young universe, it turns out that even a small percentage of metal is sufficient to cool off star-forming clouds of gas and dust enough to make the whole process more efficient. The process of the first round of Population III stars turning supernova kicked off a renaissance of star formation; lower-mass stars were able to form because the Jeans mass was much more reasonable.

Star Classification: Population II and I

Up, up, and away!

Or, in the case of Population III star explosions: Out, out, and away!

After Population III stars exploded into supernovae at the end of their lives, heavier elements were created — for the first time in the history of the universe. These elements became part of the interstellar medium, and, as such, provided a breeding ground for the next generation of stars to form. The presence of these heavier materials also had the nice consequence of making star-formation faster and more efficient. Rather than the giant hydrogen and helium puffballs of Population III, the Population II stars that formed next look much more familiar.

Stars can be classified into Population II and I based on their metal content, and scientists find that content through telescopic observations of their spectra. These observations allow us to see which lines are present or missing, due to the presence or absence of metals in their atmospheres.

Population II: Oldest observed stars formed 1 to 15 billion years ago

The initial burst of the Big Bang created hydrogen and helium, but no significant heavier metals (see Chapter 13). Our beloved planet Earth is made from heavier metals, as well as most things in the universe are — including us — and Population II stars were the next step in that journey.

TECHNICAL STUFF

Population II stars are defined as "metal-poor" because their average metallicity, or metal content, is less than 10 percent that of our Sun. Because they were birthed from an environment with more metal than Population III stars, they have a higher metal content than that first generation, but it's still not as high as stars created more recently.

Population II stars also seem to have a higher proportion of elements like oxygen and neon (produced by Type II supernovae), as compared to iron (produced by Type 1a supernovae). (See Chapter 5 for a refresher on supernova types.) This composition is different from the proportions of these elements observed in Population I stars, and might mean that Type II supernovae contributed more metals to the interstellar medium earlier in the history of the universe, when Population II stars were forming. If this is the case, then Type 1a supernovae could have become more important later, when Population I stars were forming.

Most Population II stars are anywhere from 10 to 13 billion years old — not young, by any stretch of the imagination, but younger than the Population III stars that formed soon after the Big Bang. They're also not as large as Population III stars. Scientists estimate Population II stars in the Milky Way are between 0.7 and 1 solar mass, or roughly the size of our Sun.

In our galaxy, the older Population II stars are concentrated in the Milky Way's central bulge and halo, but they can also be found in the outskirts of the disk, called the thick disk. Observations show that the stars in the halo are the lowest in metal content with $[Fe/H] < -1.0$. These stars are most often found in globular clusters. Population II stars could still form in isolated clouds of still metal-poor gas, but our galaxy seems to not have any of that kind of gas left.

TECHNICAL
STUFF

METALLICITY

The *metallicity* of a star refers to the amount of heavy elements in that star as compared to the amount of those elements in the Sun. (Remember that here, as elsewhere in astronomy, *heavy* and *metal* mean things that are more massive than hydrogen and helium.) Metallicity is defined as [X/H] where X is a particular element. To calculate the metallicity of a star, use this equation:

$$\left[\frac{X}{H}\right] = \log\left(\frac{X}{H}\right)_{star} - \log\left(\frac{X}{H}\right)_{Sun}$$

In this equation, X is the abundance of a particular element, as scaled to the abundance of hydrogen H. This means the solar metallicity, or the metallicity of a star with the same abundances as the Sun, is [X/H] = 0, by definition. You can use this measurement to compare stars easily to the Sun — iron is a typical element used for this purpose. If a star has twice as much iron as the Sun, its metallicity is [Fe/H] = 0.3, whereas a star with 10 times less iron as the Sun will have [Fe/H] = –1.0.

Population I: Young stars formed 1 million to 10 billion years ago

Just as remnants of Population III stars became a breeding ground for Population II stars, Population II stars in turn fed into the growing generation of Population I stars. The youngest stars in the classification system, Population I stars formed between 1 million and 10 billion years ago. Much like the youngest sibling in a large family gets all the hand-me-down clothes and toys, Population I stars had much more available to them in terms of stellar formation material than their predecessors.

The Sun is a Population I star with about 1.4 percent metal content, considered intermediate as compared to other stars in this population. Population I stars like the Sun are mostly located in the spiral arms of a galaxy like the Milky Way, though some older Population I stars are located in the disk of the Milky Way. These older disk stars have metallicity ranging from about [Fe/H] = -0.5 up to about [Fe/H] = 0.3, but most stars are only a bit less than the Sun with [Fe/H] = −0.2. Most open star clusters are young and consist of Population I stars, whereas globular clusters are mostly Population II objects (see Chapter 6 for more on star clusters).

The Epoch of Reionization

To recap our story of star formation in the universe:

>> Population III stars, the oldest in the universe, were forming within a billion years of the Big Bang.

>> Population II stars came into being 1 to 15 billion years ago.

>> Population I stars are the babies of the universe, forming 1 million to 10 billion years ago.

What happened after the first stars formed in the universe? Was it a smooth transition from Population III to Population II and then I? Not at all. Those early Population III stars had a big impact on the universe, in ways that are hard to even conceive of.

Here on Earth, humans impact the environment with every crop we grow, every building we construct, and every bit of clothing or plastic that we discard. Our footprint is literally everywhere, from eroding topsoil to water bottles left at the top of the world's highest mountain peaks. Every move we make affects the balance and composition of our natural environment, though the changes to that environment take place gradually enough that we may not even notice them in our lifetimes.

The same can be said about our stellar nighttime companions; when a star is born, it's not all Welcome Home cards and confetti announcements. Every star impacts the environment around it, and those effects start to add up when we're talking about the 200 billion trillion (yes, you read that right) stars that exist in the visible universe. The earliest stars had the net effect of reionizing the environment around them, ultimately allowing their light to permeate the cosmos and prepare for the dawn of the star formation era. How did this happen?

Energy bubbles and the leaching of ionized radiation

When the first stars started forming around 100 million years after the Big Bang, nuclear fusion ignited. Aside from the obvious visible effects of a star being born (hey, what's that bright light in the sky?), stars emit radiation on different wavelengths depending on the star's temperature. They can emit x-ray radiation, gamma rays, and, super-important in this context, ultraviolet (UV) radiation.

Population III stars were high in mass and low in metal. Their process of nuclear fusion was less efficient than later stars, resulting in very high surface temperatures. Some models suggest that their surface temperatures could be as high as 100,000 K (180,000°F), about 17 times hotter than the Sun. This high temperature means that most of the energy given off by these early stars was in the form of ultraviolet radiation, and it quickly heated up the surrounding gas clouds.

TECHNICAL STUFF

This heating effect led to what's called reionization of the surrounding hydrogen gas. *Ionizing* refers to the process of creating ions, or any molecule with an electric charge resulting from either gaining or losing an electron. An everyday example of ionization comes from the x-ray machine at your doctor's office. These machines use ionizing radiation to see the bones inside your body. *Reionization*, in astronomical lingo, is what happened when the atoms in the neutral hydrogen gas started discarding electrons; the ultraviolet light from these first stars split the hydrogen atoms into separate electrons and protons.

Bubbles of ionized gas surrounding these first stars grew and created holes in the neutral cosmic gas, eventually merging as more stars formed over the first few hundred million years after the Big Bang. This process continued until the gas between the newly forming stars and galaxies was completely ionized. The net effect of these ultraviolet emissions effectively reionized the universe, hence the name *Epoch of Reionization*; this epoch dates to about 600 million years after the Big Bang. How do we know this age? The James Webb Space Telescope is revealing early galaxies at this age, and distant quasars from 1 billion years after the Big Bang show strong evidence for absorptions of UV light, probably from this reionization.

Let there be light

So, to summarize what we know here, take a look at Figure 14-2 for a timeline of events following the Big Bang. We start off with a Bang (sorry, couldn't resist) and then the first round of ionization takes place, leading up to the emission of the cosmic microwave background radiation (see Chapter 13) about 380,000 years after the Big Bang. This radiation, detectable here on Earth, is sometimes called the *afterglow* of the Big Bang.

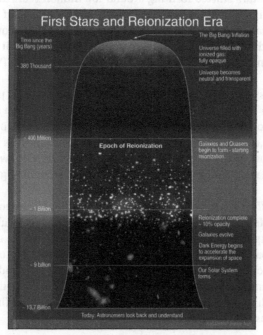

FIGURE 14-2: First stars and the Reionization Era.

Courtesy of NASA/WMAP Science Team / Public Domain

The universe subsequently began to form stars, and eventually galaxies, from about 300 million years post–Big Bang leading up to the period of reionization about 600 million years after the Big Bang. At that point, the reionization was complete.

TECHNICAL STUFF

Why is this reionization so important? Reionization indicates the end of the long period of Cosmic Dark Ages, during which the universe was largely filled with neutral hydrogen. When stars finally began forming, detectable light was given off by the newly formed stars, but most of it was trapped in that sea of hydrogen gas. After most of the universe was reionized, light across much of the electromagnetic spectrum's range could travel through the cosmos. This point marks the transition where the universe cast off its primordial haze, allowing future humans (us) to experience and study the cosmos. Sensitive telescopes like the James Webb Space Telescope are currently studying the nature of reionization by observing faint, distant objects at high redshifts in the infrared, effectively paving the way to study the very first stars.

Formation of the First Galaxies

You can't make an omelet without breaking a few eggs, and you can't have a galaxy without stars. It took time after the Big Bang for galaxies to form because in those first seconds, minutes, and years after the Big Bang, stars hadn't formed yet. No stars = no galaxies. Fast forward about a few hundred million years after the Big Bang, and we now have stars (refer to Figure 14-2). What next?

Opposites (or not) attract!

TIP

It's a bit of a chicken-and-egg question whether the first galaxies formed from clumps of matter and then formed stars inside them, or stars formed first and then clumped together into the first galaxies. Whichever galaxy formation theory you follow, it's the same: After enough stars were formed, the world's first cosmic speed-dating event took place. Instead of poor lighting and bad music, the scene was set with gravitational attraction, and lots of it. As more stars became gravitationally attracted to each other, they formed into galaxies; the dust and gas left over from star-formation eventually became the predecessors to nebulae.

WARNING

The first galaxies would not have been the perfectly spherical models that you may have seen in the movies. They were likely irregular, clumpy affairs, and their initial structure was likely driven by dark matter. From what we know about dark matter, smaller objects such as minor galaxies tended to form first; these smaller objects are then drawn together into larger ones. Whether the stars-first or the galaxies-first theory proves correct, astrophysicists think that the earliest galaxies then started to merge to become more and more massive.

TECHNICAL STUFF

And how old is the oldest galaxy, you ask? The oldest galaxy captured so far via the JWST is GLASS-z13. It's visible using colored wavelengths of infrared light to estimate distances via redshift. The scientific team behind the discovery estimated its age at 13.5 billion years — only 300 million years after the Big Bang and the origin of the universe. Massive would be an understatement; GLASS-z13 is estimated to have a solar mass of 1 billion solar masses. That's one billion times as massive as our Sun.

Continued evolution and creation of new galaxies

Many things in life are unavoidable. These include stubbing your toe on the edge of the bedframe, finding toys in your shoes, or stepping out to run an errand the minute before a mail carrier appears with a Signature Required package. On a cosmic level, the process of creating new galaxies is equally unavoidable. Time stops for no one, and neither does galaxy creation. As the universe evolves, new galaxies are created and smaller galaxies merge together.

WARNING

As you learn in Chapter 10, the process of galaxy mergers continues to the current day; one day in the distant future, it will even happen to our own home as the Milky Way and Andromeda galaxies become one.

Telescopes like the James Webb Space Telescope are our best bet to observe this cosmic history, helping us understand the past and perhaps better predict the future. Even though the light emitted from stars and galaxies in the distant past was mostly in visible and ultraviolet wavelengths, this light is redshifted as it travels to us through time and space due to the expanding universe. By the time it reaches Earth, light from the early days of the universe is now visible at infrared wavelengths — exactly what JWST is designed to study. Stay tuned for more new observations from the newest great space observatory.

Chapter 15

And Then It Gets Weirder: Dark Matter, Dark Energy, and Relativity

lthough there are numerous origin stories of the yin-yang symbol, ancient Chinese philosophy generally describes the concept of yin and yang as the union of opposing forces, or Heaven and Earth coming together in counterbalance with each other. Consider the Big Bang and its aftermath from this perspective. The Big Bang caused all space in the universe to want to expand and spread out the mass-energy of the universe into a growing volume, and this expansion was almost immediately countered by the gravitational force trying to pull it all back in and collapse into itself. Understanding relativity is a heroic task for anyone, but it's time to put on your superhero cape. In this chapter, we look at how Einstein's theories explain the ways in which space and time were united and put you on the starting block in the race to map out the origins of the universe.

Although most aspects of our origin story have been battle-tested by astrophysicists for generations, recent discoveries highlight the need to introduce components that humans cannot see, detect, or observe directly. Uncovering the origins and makeup of our universe requires astrophysicists to be detectives. They use multi-wavelength observations, techniques, and special instruments to decode space and retell the story of the universe, but science proves that certain aspects of this story can only be detected by their absence. Cut to stage left: Dark energy and dark matter! These concepts are critical to understanding the nature of the universe but are also only detectable currently by their influence on other parts of the cosmos. Make that superhero cape an invisibility cloak, and you're well on your way.

General Facts about General Relativity

Though Einstein took a somewhat meandering path to general relativity, his seminal theory paved the way for everything we understand about the cosmos today, from galaxies and nebulae to singularities and black holes. Start with German-American physicist Albert Einstein's theory of special relativity from 1905. This landmark theory famously generated the equation relating matter (m), energy (E), and the speed of light (c) into $E=mc^2$. While contemplating Newton's laws of gravitation, Einstein struggled for years with a way to make sense of his perception that astronomical bodies did not push each other away, but instead attracted each other. His concept of general relativity was based not only on astronomical observation and physics, but on mathematical models that combined the forces of nature with the math that made his observations possible.

REMEMBER

Before we dig into general relativity, remember that Einstein's predictions were, at the time, unproven; they were pure theory. In the last hundred years since general relativity was published, virtually every prediction in his theories has been confirmed when technology advanced to the point of being able to prove them.

Keeping it special

Before you can understand general relativity, you have to start with a special case: special relativity. Remember Newton's laws of motion, (see Chapter 2)? Those laws, developed back in the 1600s, work well for most regular scenarios you encounter here on Earth. Objects in motion tend to stay in motion, forces produce an acceleration, and bodies attract each other with a force depending on their masses and the distance between them.

These laws largely held true for hundreds of years but were not without their problems. One of the first problematic areas was light. Newton's laws suggested that light, as a wave, had to move through some kind of matter. Scientists theorized that a mysterious "ether" might have existed between the stars, but one was never found. Eventually physicists determined that light could move through a vacuum at the fixed speed of 186,000 miles per second (300,000 kilometers per second). Light can also move through air and glass and other things . . . it just moves slower in non-vacuums.

Through a series of thought experiments mostly involving trains, a young Einstein figured out that because the speed of light is a constant, time itself must be relative. Put another way, time moves differently depending on how an observer is moving, and this speed can be slower than how time moves for a stationary observer. This concept results in all sorts of strange implications such as time dilation, the idea that time passes more slowly for a fast-moving object than it does for one that's not moving.

The time dilation effect is almost negligible at normal Earth-related speeds, but if you start traveling at speeds close to the speed of light (for example, on a space-ship), time would move much more slowly for you than it would for those you left behind. That vacation to Alpha Centauri might mean that everyone you know would be long dead before you return to Earth! This scenario isn't one to lose sleep over because engineers have yet to invent ships that can travel anywhere near the speed of light, but there's the potential for larger effects on future space-faring civilizations.

A corollary to time dilation is length contraction. Because the speed of light is always a constant regardless of reference frame, time and space themselves must change depending on how fast you're moving. If that's not enough, another important consequence of special relativity is the famous equation $E=mc^2$. This relationship shows that matter and energy have an equivalence, though because c is the speed of light, the amount of energy released by turning even a small amount of matter into pure energy would be huge.

Einstein's explanation of gravity's interaction with space-time

Einstein published his theory of special relativity in 1905, and despite its publicity and success, Einstein wasn't satisfied. He spent the next ten years developing the theory of general relativity, one that expands special relativity to include gravity. In a nutshell, Einstein's discovery here was that massive objects warp and bend space-time using the force of gravity.

TIP

Although it's virtually impossible to simplify general relativity, you can break it down to the essential elements. Einstein created a relationship between the three dimensions of space and the fourth dimension of time such that space and time interact in a model called space-time. The interactions between all four dimensions can be defined by the Einstein field equations, a mathematical model for how space-time is curved by matter. As a result of this theory, the gravitational field is essentially a consequence of curved space-time.

Einstein's field equations describe

>> The way in which gravity is a by-product of space-time curvature

>> The relationship between momentum and energy for any point in time and space

The actual field equations can be written in a deceptively simple form:

$$G_{\mu\nu} \equiv R_{\mu\nu} - \frac{1}{2} R g_{\mu\nu} = \kappa T_{\mu\nu}$$

In this equation, $G_{\mu\nu}$ is the Einstein tensor, $R_{\mu\nu}$ is the Ricci tensor, R and g relate to the curvature scalar, κ is a constant, and $T_{\mu\nu}$ is the energy-momentum tensor.

REMEMBER

What does this all mean? Back in Newtonian gravity, the source of gravity was mass. However, in relativity, mass becomes part of a generalized energy-momentum tensor (a three-dimensional object). This tensor includes effects like energy, momentum, pressure, and shear stress, and it works well (mathematically speaking) in curved space-time. The gravitational field equation relates the energy-momentum tensor to one called the Ricci tensor, one that demonstrates the effect of gravity on a cloud of test particles that are allowed to fall freely under the influence of gravity from an initial resting position.

TIP

Ideas like the conservation of the energy-momentum tensor can also be extended to curved space-time by using a lot of math and complicated geometry, including curved manifolds and covariant derivatives. These equations can be related back to Newtonian mechanics, which is a limiting case where the local gravity field is weak and/or velocities are slow (much less than the speed of light). In this case, the constant of proportionality κ is

$$\kappa = \frac{8\pi G}{c^4}$$

where G is the Newtonian gravitational constant from earlier and c is the speed of light.

SPECIAL RELATIVITY VS. GENERAL RELATIVITY

Einstein created two theories of relativity: special and general. Here are the differences in brief:

Special relativity: 1905, the interconnected nature of energy, mass, space, and time:

- The speed of light in a vacuum is a constant, no matter where you are or where that light is going.

- Motion of anything in space is relative to the motion of everything else in space.

- The laws of physics appear the same anywhere in the universe as long as you are not accelerating. (They are the same, but if you're accelerating as you fall from an airplane, the apple you drop won't exactly appear to be affected by gravity.)

- The speed of light defines a relationship with matter and energy such that $E=mc^2$.

General relativity: 1916, the integration of gravity:

- Time and space are aspects of the space-time continuum.

- Energy and matter have the effect of curving space-time via gravity.

- Serves to unite gravity with space-time.

The theory of general relativity followed special relativity, mainly focused on how the speed of light determines a relationship between matter and energy. Special relativity had no specific concern with gravity, the key factor in general relativity.

Space-time curvature and total forces

General relativity establishes space-time as a four-dimensional construct. But why does it curve? The answer, as Einstein explains it, comes from large, massive bodies that distort the space-time fabric according to their mass. Per the Einstein equation above, space-time curvature exists when matter, momentum, and energy coexist. Check out Figure 12-3 for a visualization of how a massive object, like Earth, could warp space-time — but in reality, the warping would be in three dimensions, not just two!

The forces from Newtonian gravity and general relativity can work together to give you the total force on an object that is orbiting around a massive central body. This total force can be written as

$$F_f(r) = -\frac{GMm}{r^2} + \frac{L^2}{mr^3} - \frac{3GML^2}{mc^2r^4}$$

TECHNICAL STUFF

In this equation G is the Newtonian gravitational constant, M is the mass of the large central object, and m is the mass of the orbiting object. R is the orbital distance, L is the angular momentum, and c is the speed of light. The first term is the force due to Newtonian gravity, and should look familiar from earlier examples. The second term is the centrifugal force on the object due to its circular motion, and the third term is the contribution due to relativity (a clue here is the speed of light on the bottom of this term). This equation shows that because the massive speed of light is on the bottom of the third term, the third term is very small and can be ignored . . . unless the mass of the central object is huge, the velocity and therefore angular momentum of the orbiting object is huge, or both.

The three tests for general relativity

As part of his theory of general relativity, Einstein proposed three tests that would serve as verification. These are

>> Gravitational redshift

>> Perihelion precession of Mercury

>> Deflection of sunlight

Well, that was a mouthful! Let's break it down. Figure 15-2 illustrates these three concepts.

TIP

Gravitational redshift is an effect in which the light given off by a massive object, such as a galaxy, will be shifted to slightly longer wavelengths due to the mass of the galaxy itself. This additional source of redshift needs to be added to the original redshift from both the motion of the galaxy and the redshift due to the expansion of the universe, where the gravitational redshift is the smallest of the three. This effect was first detected 43 years after Einstein predicted it in a laboratory setting using gamma rays, but it was much harder to measure in an astronomical context.

Recent studies have performed a statistical redshift analysis of huge numbers of galaxies and found that an additional redshift term was needed that exactly matched Einstein's predictions for gravitational redshift due to general relativity. This additional term? You guessed it: gravitational redshift. Evidence of gravitational redshift has also recently been found using x-ray spectra from the Chandra X-ray Observatory of a binary star system that includes a neutron star. Spectral lines of iron and silicon are redshifted in a way consistent with a gravitational redshift from the neutron star's immense gravity, as shown in Figure 15-1 for iron.

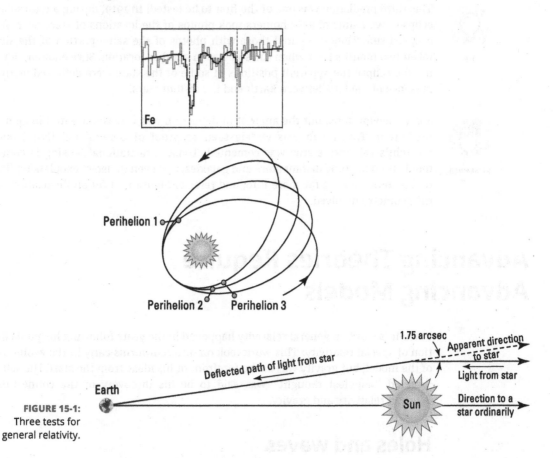

FIGURE 15-1:
Three tests for
general relativity.

The second prediction involves the position of Mercury over time during its orbit around the Sun. Because Mercury is so close to the Sun, the Newtonian theory of gravity predicted that its orbital alignment should change — precess — over time. Frustratingly, as observations and timekeeping got better, it was realized the precession of Mercury's orbit didn't match Newton's calculations. However, under general relativity, Einstein predicted that the orbit should precess more than was predicted using Newton's gravity because the Sun warps space-time near Mercury's orbit. Bingo! A discrepancy between calculations and reality was solved. This effect is illustrated in Figure 15-2.

Einstein's third prediction was that gravity could actually bend light from stars or galaxies. This effect was predicted by Newtonian gravity; under general relativity, Einstein calculated that light would be bent roughly double that of previous expectations.

TIP

This third prediction was one of the first to be tested! In 1919, during a total solar eclipse, two teams of astronomers took photos of the locations of stars near the eclipsed sun. They compared those with photos of the same portion of the sky taken two months later when the sun was in another location. Sure enough, during the eclipse the apparent positions of some of the stars were deflected by the presence of the Sun between Earth and the distant stars.

REMEMBER

Sorry, Newton fans, but the amount of deflection was consistent with Einstein's prediction. This result was widely seen as proof of General Relativity, and Einstein's celebrity status was cemented. Today, gravitational lensing is commonly used to study distant stars and galaxies, and even to detect exoplanets; flip to Chapters 7 and 12 for more info, and refer to Figure 15-1 for an illustration of the geometry involved.

Advancing Theories Require Advancing Models

Einstein's work on general relativity happened in the years following his publication of special relativity. This work took on different forms early in the evolution of the model, but gravity was at the forefront of his ideas from the start. His self-declared "happiest thought" was said to be his discovery of the connection between relativity and gravity.

Holes and waves

Soon after the publication of general relativity, other astronomers and physicists started grappling with the implications of this new mathematical model of the universe. Some of the implications and further results of general relativity include

>> **Black holes:** German physicist Karl Schwarzchild found a solution to the Einstein field equations that reduced down to a singularity, leading to the prediction of black holes (see Chapter 8).

>> **Gravitational waves:** Distortions in the fabric of space-time, predicted by Einstein, were finally detected in 2015 (See Chapter 8).

First general relativity models: A stable universe and its challengers

Soon after his publication of general relativity, Einstein turned his mind to contemplating the implications of relativity for the universe as a whole. Based on the assumptions of the day, Einstein took a starting point from the idea that the universe was static and unchanging in time. He also included a few other assumptions. One was that all reference frames are equal, that is, the laws of physics are the same anywhere in the universe. A second assumption was that the inertia of a body, or the resistance of that body to movement, is due to the presence of other, external masses.

In a 1917 paper, Einstein showed that a model of the static universe could be consistent with relativity, but that it required two further assumptions:

» The universe needed to be closed in both space and time, without a beginning or end.

» A constant, which Einstein called the "Cosmological Constant (Λ)," needed to be added to the field equations to balance things out.

Einstein's solution seemed self-consistent, and it also predicted that there would be a linked relationship between density of matter and the size of the universe (although, as it later turned out, if the density of matter varied, the universe would be unstable — not a good thing, in terms of our continued existence here). As is typical in science, the publishing of Einstein's paper brought detractors out of the woodwork. Other scientists began working through the implications of his findings and, in some cases, trying to poke holes in Einstein's interpretations.

A serious challenge to Einstein's version of cosmology came from Russian physicist Alexander Friedman. Friedman suggested that non-static models needed to be considered as solutions to the Einstein Field equations. Friedman found an elegant set of solutions to the field equations that showed how the time evolution of the universe could be related to changes in the density of matter and the cosmological constant.

TECHNICAL STUFF

Another challenge came from Georges Lemaitre, a Belgian physicist who found a series of solutions to Einstein's modified field equations that allowed the radius of the universe to vary with time. Lemaitre suggested that this was evidence that the universe could be expanding (see Chapter 13 for more on the implications of Lemaitre's solutions).

Einstein's "greatest mistake" and its reinterpretation

The theoretical challenges to Einstein's view of cosmology, including the Cosmological Constant and the assumption of a static universe, didn't convince Einstein that his idea was invalid. What did convince him, though, was observational evidence that provided sufficient proof to tip him over to a different view.

TECHNICAL
STUFF

Hubble's evidence for an expanding universe was extremely persuasive. Hubble took measurements of the redshift of galaxies, proving that there was a relationship between the distance to a galaxy and the speed at which it was moving away from us (see Chapter 9 for more on Hubble's Law). When Hubble's results were published in 1929, Einstein converted over to the idea that the universe was expanding and could vary over time, even though this idea conflicted with his own static view. Einstein published new models that included the expanding universe, and eventually abandoned the Cosmological Constant as no longer necessary. Looking back, Einstein is even said to have referred to the Cosmological Constant as one of his biggest mistakes.

TECHNICAL
STUFF

If you think we're done with the story of the cosmological constant, think again. A major area of ongoing research in cosmology is an attempt to unite Einstein's theory of general relativity with the theory of quantum mechanics, a statistical description of the behavior of the universe on the very smallest scale. The Einstein field equations of general relativity may actually need an extra term, currently interpreted as the energy density of space. This term is also known as vacuum energy, which is an important concept both in quantum mechanics and in dark energy. More on that later in this chapter.

PARALLEL UNIVERSES AND THE MULTIVERSE

Have you ever watched a television show in which a character meets their counterpart from a parallel universe, where things are mostly the same with a few subtle differences (mustaches, non-existence of doughnuts, a different outcome for WWII)? Could this scenario ever be possible in real life? Some theoretical physicists say yes — and it could be an outcome of general relativity!

One of the biggest problems right now in theoretical physics is the attempt to reconcile general relativity (does a great job explaining gravity and star or galaxy-scale effects in the universe) with quantum mechanics (does a great job of explaining tiny atom-scale effects and can explain the united strong, weak, and electromagnetic forces). Theoretical physicists try to make these two ideas meet in the middle, and one of the most promising attempts is string theory. To vastly oversimplify string theory, the main idea is that tiny strings vibrating in 10 or 11 dimensions could connect the large and the small scales and make quantum mechanics work with general relativity.

And what about parallel universes? String theory has more than those tiny vibrating strings. There are also large two-dimensional surfaces called membranes. It's possible that our entire universe could be encapsulated on the surface of one of these membranes, and there could also be other surfaces out there. These other surfaces may contain universes where the laws of physics are completely different and bizarre, universes where it's impossible for life (or even matter) to exist. So why are the laws of physics in our own universe optimized for life? It could be that of all those universes out there in the multiverse, there are only a few where life could exist. Let's just be glad that we are in one!

Galactic Glue: Dark Matter

TIP

If the puzzle we've laid out so far is still missing a few pieces, dark matter will fill in at least one of the gaps. There's more than meets the eye in the universe, and we mean that literally. Observation has produced clear evidence that the universe contains more mass than we can see. The general name given to this type of unknown mass is *dark matter*. What we do know is that dark matter is, as the name suggests, matter — it's made up of particles, but ones that don't emit, reflect, or absorb light.

So why do we think dark matter exists if we can't see it? Put your astronomical detective cap back on and look for clues — which is, by the way, exactly what astrophysicists do when analyzing images and data. Dark matter has clear effects on other objects that we can see and observe, and scientists can therefore observe and derive data about dark matter indirectly. We talk about this in Chapter 11, where observations of the dynamics of galaxy clusters require a large percentage of unseen, in other words, dark, matter.

TECHNICAL STUFF

One of the big reasons to study dark matter is that it's not a tiny part of the universe. On the contrary — take a look at Figure 15-2. The stars, planets, and other visible objects in the cosmos are thought to comprise only about 5 percent of total matter. Around 68 percent is thought to be dark energy (more on this in the section "Detecting the darkness"), leaving about 27 percent of the universe as dark matter. That's more than five times as much dark matter as there is visible matter! And what does it take to break through the darkness?

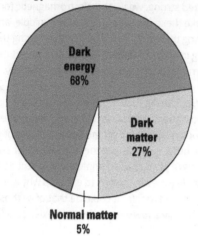

Energy distribution of the universe

Dark energy 68%

Dark matter 27%

Normal matter 5%

FIGURE 15-2:
Types of energy
in the universe.

Detecting the darkness

Dark matter is hard to define precisely because it can't be seen, touched, or measured. However, astrophysicists have a pretty good handle on what it isn't. Dark matter is not, for example, a planet or a star.

TECHNICAL
STUFF

Dark matter is not antimatter, which is essentially matter with inverted electrical charges. It's not composed of detectable matter such as baryons, and it's certainly not visible like your best friends, dust and gas. Lack of sufficient gravitational lensing also tells scientists that dark matter isn't just a really big black hole.

REMEMBER

How do we know dark matter is there? Because you can't see it directly, the only choice is to look indirectly for its effects on regular — that is, luminous — matter. In Chapter 11, we talk about how in the 1930s, German astronomer Fritz Zwicky used a clever method of measuring the mass of a nearby galaxy cluster. However, when he compared the measured mass to his estimate based on the total brightness of the cluster, he found that the galaxy cluster was 200 to 500 times less bright than it should be, compared to the total measured mass. The best explanation for this is that the galaxy cluster contains large amounts of dark matter, providing mass but not adding to the total brightness.

Later, in the 1970s, American astronomers Vera Rubin and Kent Ford also measured the rotational speed of individual galaxies. They learned that an extra source of mass was needed to provide enough gravity to keep them from breaking apart due to their rotation rates. Sure enough, dark matter again! See Chapter 10 for more on this discovery.

DARK MATTER ALTERNATIVES

Are there any alternatives to dark matter? From what astronomers have seen so far of the universe, either

- Dark matter exists, or

- The laws of physics as we know them, including Newton's Laws and even general relativity, are incomplete when you look at scales as big as galaxies.

Some cosmologists study an idea dubbed *modified gravity*, but it's proven difficult to find a theory of modified gravity that actually holds together — no pun intended! Such a theory would have to explain all the places where dark matter seems to fill the gaps — from galaxy rotation and galaxy cluster masses, all the way to imprints that we can see from sound waves in the early universe. So far, no theory of modified gravity has been discovered that works to explain all these observations. That said, it's possible there is both dark matter and a modification needed to gravity. Data from the Vera Rubin Observatory, currently under construction in Chile, should improve our understanding in the next few years.

REMEMBER

Gravitational lensing is the bending of light by massive objects (see Chapter 12). This technique can also be used to estimate the amount of mass in between us and a lensed object. Figure 15-3 shows a Hubble image of galaxy cluster CI 0024+17 — the arcs on the left image show the gravitationally lensed images of galaxies that are located far behind the galaxy cluster. The fuzzy portions of the image on the right show where dark matter must be located in the galaxy cluster in order to produce the observed lensing effect.

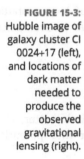

FIGURE 15-3:
Hubble image of galaxy cluster CI 0024+17 (left), and locations of dark matter needed to produce the observed gravitational lensing (right).

Courtesy of NASA, ESA, M.J. Jee and H. Ford (Johns Hopkins University)

Don't be a WIMP(s): Weakly interacting massive particles

Okay, so there's good evidence that dark matter exists. But what is it? People come in all shapes and sizes, so do plants, stars, and virtually everything else in existence. Matter is no different. It's thought that on a universal scale, there's at least two types of matter:

>> Baryonic matter composed of familiar neutrons and protons (includes planets, stars, and more)

>> Dark matter

One theory on dark matter composition is on the wimpy side. And by that, we mean that dark matter may be composed of weakly interacting massive particles, or WIMPs. Two other possibilities are axions or sterile neutrinos.

TECHNICAL STUFF

Axions are a hypothetical low-energy, low-mass particle related to the so-called *strong force*, one of the four fundamental forces discussed in Chapter 2. The strong force acts on subatomic particles by holding them together, in particular as it binds quarks and other particles into atomic nuclei. Dark matter could be composed of axions, or, on the other end of the mass spectrum, it could also be composed of WIMPS. WIMPs are similar to axions but interact through the nuclear weak force instead of the strong force. So far, none of the predicted WIMPs have been found, but the theory isn't quite dead yet.

Other possible dark matter candidates are sterile neutrinos. Similar to normal neutrinos, these are fundamental particles that have a small amount of mass, no charge, and a very low probability of interacting with other particles. Although previously detected neutrinos all have some kind of a spin, sterile neutrinos have no spin and likely only interact with gravity.

MACHOs, massive compact halo objects, are a source of noise in our understanding of non-baryonic dark matter. They are made of normal baryonic matter that's non-luminous. They may be composed of brown dwarf stars, black dwarfs, or stellar black holes, or they could even consist of isolated chunks of dense metallic material. MACHOs act like dark matter but aren't the dark matter we seek since they are actually made out of normal matter.

Some like it hot, some like it cold

Within the large and somewhat nonspecific characterization of dark matter, scientists have found it easier to start classifying dark matter into potential categories based on temperature. Scientists theorize that dark matter can be "cold," "warm," or "hot." The temperature designation here relates mostly to the speed of the dark matter particles as they moved on their merry way through the early universe, soon after the Big Bang. Their exact speed would have depended on the mass of those particles, and the temperature designation would reflect how hot the surrounding universe was at the moment that dark matter particle came into being.

>> Hot dark matter consists of light, quick-moving particles — relic neutrinos from the early universe are hot.

>> Warm dark matter is an in-between category — sterile neutrinos would be warm.

>> Cold dark matter's particles are heavier and move much slower — WIMPS would be cold.

Why are these categories important? If a dark matter particle is hot, meaning light and fast, it can move quickly and will tend to smooth out any structures. Alternatively, cold dark matter particles, which are heavier and move more slowly, will tend to build up structures in the universe.

TECHNICAL
STUFF

Astrophysicists think that in the early universe, soon after the Big Bang when the first particles and atomic nuclei were forming, dark matter (possibly of multiple types) was also created in high-energy collisions during this high-density, high-energy time. The dark matter streamed out through the universe, interacting and shaping the growing structures. Cold dark matter such as WIMPs would have helped stick together and perhaps attract normal baryonic matter, in the process aiding the later creation of stars and galaxies. Hot dark matter, such as relic neutrinos, would have moved quickly through the early universe and smoothed out any forming higher-density areas. The structure we see in the universe is evidence that hot dark matter, if it existed at all, was a much more minor component than cold dark matter. Warm dark matter (sterile neutrinos) is still a theoretical possibility and could have assisted in the building of early structures in the universe.

And of course, when you have categories you also need particles that defy characterization! In this case, axions are both light and extremely cold and don't fit neatly into one of these categories. They also interact with normal matter very weakly, and their temperature is so close to absolute zero that they are barely moving as compared to other types of matter. Although axions may exist, they also are unlikely to have contributed much to the building of large-scale structures in the universe.

Dark Energy in Review

Thinking back to Figure 15-3, remember that 5 percent of the universe is made of ordinary baryonic matter (people, turtles, stars, and so on). Twenty-seven percent is composed of dark matter — weird stuff, for sure, but we have some solid evidence that it exists and interacts gravitationally with our nice normal matter. And what about that other 68 percent of the universe? That's where it gets even weirder, and where dark energy comes into play.

Why do we need dark energy?

So why do we need dark energy? Remember those important studies from the 1920s by Edwin Hubble, ideas that solidified the theory that the universe is constantly expanding. For most of the 20th century, astrophysicists were confident that they understood the timeline. The essential story was that the Big Bang happened and the universe started expanding, but over time, billions and billions of years, the gravitational attraction of all the matter in the universe would be enough to slow that expansion down.

That theory changed significantly in 1998, when observations with the Hubble Space Telescope of extremely distant supernovae showed that way back when these supernovae happened, the universe was actually expanding more slowly than it is today. This means that contrary to popular belief, the expansion of the universe was actually speeding up, not slowing down due to the force of gravity! Figure 15-4 shows one of these ancient supernovae, SN 1997ff, as seen by the Hubble Space Telescope — the bottom right image is a different image that shows a new bright "star" appearing between an observation in 1995 and one in 1997. That's SN 1997ff, which is about 10 billion years old.

So how could the universe's rate of expansion be increasing over time rather than decreasing? An undetermined force must be responsible for that increased rate of expansion. This potential solution came to be known as dark energy, and it's the best explanation we currently have as to why the galaxy's expansion is accelerating.

Are dark matter and dark energy the same? Confusingly, no — they are quite different. Dark matter is the name for that invisible substance responsible for keeping galaxies together, whereas dark energy is the force that pushes the universe apart. How's that for polar opposites?

FIGURE 15-4:
Ancient Super-
nova 1997ff, as
seen by the
Hubble Space
Telescope.

Distant Supernova in the Hubble Deep Field
Hubble Space Telescope • WFPC2

NASA and A. Riess (STScI) • STScI-PRC01-09

Courtesy of NASA and Adam Riess (STScI)

A story of accelerating expansion

REMEMBER

In the Big Bang discussion in Chapter 13, you learn that the universe began with an explosion at a singularity, and then started expanding and cooling. Key point: The universe began expanding and creating space right after the Big Bang. The fundamental reason for the slowing down of this expansion is the gravitational attraction of matter as predicted by Einstein's theory of general relativity.

Astrophysicists now think that there was a transition period around 9 billion years ago. Dark energy began dominating the universe; at this point, the expansion of the universe started accelerating instead of decelerating. The observation of supernova 1997ff, as discussed in the earlier section "Why do we need dark energy," is from roughly 10 billion years ago. This supernova predates the transition period when the acceleration of the universe began to slow. The dating here is key, and it explains why the measured acceleration from 1997ff is slower than the measured acceleration of more recent astronomical objects (ones that are closer to us). Figure 15-5 shows a diagram of the expanding universe, with a change in expansion rate at about 9 billion years ago.

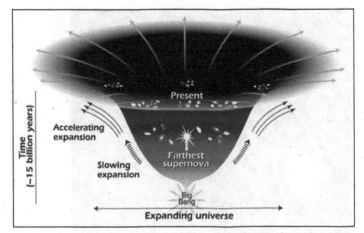

FIGURE 15-5:
Diagram showing changes in expansion of the universe due to dark energy.

Courtesy of NASA/ STScI/ Ann Field / Public Domain

This shift between decelerating and accelerating expansion is relatively *(ha!)* solid proof for the existence of dark energy. Astronomers using the Hubble Space Telescope measured dozens of ancient supernovae in an effort to continually refine our knowledge of cosmic expansion rates, and they are starting to be able to map out approximately when this transition took place. One possibility for the timing of this transition is that it occurred when the early universe's matter spread out enough that gravity no longer had the upper hand. After the universe had grown large enough, dark energy could take over. On the smaller length scales common in our solar system and galaxy, gravity would still dominate; at these huge length scales, dark energy would be the more important force.

Okay, so dark energy seems to be necessary. But what is it? Good question!

Where Did Dark Energy Come From?

Energy comes in many shapes and forms. At a very high level, in the realm of traditional Earth-based physics, energy can be one of two types:

>> Kinetic (the energy of motion)

>> Potential (stored energy)

Energy can be produced or generated by different means: A few include gravitational energy, thermal energy, electrical energy, or nuclear energy.

Dark energy, on the other hand, has no simple counterpart or familiar analogy. It makes up around 68 percent of the universe and is responsible for the accelerating rate of expansion of the universe. Beyond that, though, what is dark energy and where does it come from?

Astrophysicists have come up with a number of possible theories for this mysterious repulsive force. Here are five of the top current ideas:

>> Empty space has vacuum energy

>> Space has virtual particles as a result of quantum field theory

>> Space is filled with an undetectable energy that acts like a pressure

>> Dark energy is created by faster-than-light particles called tachyons

>> The treatment of gravity in general relativity needs an extra term

These might all seem like strange explanations. To be honest, you're officially now at the point where astrophysics gets mysterious (even more mysterious than it already was). Being scientists, though, astrophysicists do have some evidence for many of these theories and have also worked to pull the idea of dark energy together with the rest of our understanding of the universe into a standard cosmological model. Here's what we do know.

Origins of dark energy 1: Einstein (again!) and the Cosmological Constant

Einstein included a term to balance out the universe and make it static; this term was called the Cosmological Constant. (The constant lets us describe how dark energy affects the universe but doesn't tell us what causes it, sadly.) Although that term was quickly discarded after Hubble's evidence of an expanding universe became known, the idea of this constant has recently been coming back as an energy density of empty space. Even empty space has its own intrinsic properties and perhaps isn't so empty after all.

TIP

One of the all-time great thought experiments is the vacuum. Defined in physics as empty space, or space without matter, anything existing in a vacuum would make that vacuum cease to exist because it would no longer be a vacuum. Space is close to being a vacuum because it is, by definition, matter-free and nearly pressure-free; it is not, however, a perfect vacuum because even in its uncluttered areas, bits of gas and dust can be found drifting — and maybe something else is there too.

How could a vacuum relate to dark energy? In developing the theory of general relativity, Einstein posited that one possible source of a universal repulsive force was space itself. Just as space is not quite a vacuum, empty space is also not quite empty, and Einstein realized that space is able to make way for more space to be created. Looking at this with modern eyes, we can propose that a form of energy such as dark energy would make this scenario possible. It's possible that dark energy takes the form of the Cosmological Constant, and is a term representing an energy that seeps out of the vacuum of space itself. This concept is sometimes known as a vacuum energy density — meaning that even in a vacuum, space itself has some intrinsic properties, including its own energy.

Origins of dark energy 2: Quantum theory

Still not convinced? Quantum mechanics would like to have a turn at trying to explain dark energy. Try this one on for size.

TIP

The quantum theory of matter posits that all matter essentially has a split personality; it has both wave and particle properties. Standard physics covers the nature of particles thanks to Newton's laws, and the nature of waves is explained by quantum physics. The idea behind this theory is that the wave and particle sides of matter create interim particles that appear, then disappear.

Under this quantum theory of matter, empty space is filled with temporary particles that, according to quantum mechanics, form and disappear constantly as "virtual particles." The existence of these particles could explain how empty space may acquire energy as a consequence of their brief existence.

TECHNICAL STUFF

Unfortunately, there's at least one problem with this theory. Based on this theoretical explanation, astrophysicists have been able to calculate how much energy empty space should contain based on the volume of space in a particular galaxy and its expected luminous mass based on its brightness. The result was concerning because per the calculations, this theory would give empty space far too much energy — 10^{120} times more than can be explained by the observations of supernovae redshifts. Although an order of magnitude or two in difference is to be expected with theoretical calculations of this nature, such a huge discrepancy is emblematic of a problem with the theory.

Quantum field theory is still working to try to resolve this huge mismatch. Perhaps a balance is lurking such that individual terms are huge but their sum is closer to zero. Or maybe the actual value of zero-point energy is randomly generated through quantum fluctuations, and we can only detect it in locations, like our universe, where its value allows the existence of normal matter.

Origins of dark energy 3: Quintessence

If space producing itself through dark energy doesn't quite convince you, another theory as to where dark energy comes from lies in the idea of *quintessence*. This as-yet hypothetical construct creates a form of dark energy with a repellant (also called *repulsive*, though *repellant* sounds a little more friendly) gravitational force that serves to increase the universe's rate of expansion.

In this theory, space is filled with an undetectable energy field that acts like a dynamical fluid that produces pressure. This material was called quintessence, after the fifth element defined in ancient Greek philosophy. To the ancient Greeks, the four major elements were fire, earth, air, and water. Aristotle added the idea of *aether* as a fifth element to represent what stars were made out of, and it was also called *quintessence*, which comes from the root quint for five. Of course, Aristotle would perhaps have been surprised to find out that the stars actually are made up of earthly materials like hydrogen and iron, but his idea of a fifth celestial element was reborn in this new idea of a property that fills empty space.

The theory is that quintessence would fill space but would also cause the expansion of the universe to accelerate, behavior fundamentally opposite from normal energy or matter. Despite the spiffy name, we still have no idea what this material really is, how it interacts with regular matter, or how to detect it — or even if it truly is a valid explanation for the existence of dark energy.

Origins of dark energy 4: Tachyons moving faster than light

Another related possible theory is that empty space is not so empty, but rather occupied by a field made of tachyons. These are hypothetical particles that can actually travel faster than the speed of light, meaning they could even travel back in time. Don't run out to the tachyon just yet, though — this theory is almost impossible to prove, and here's why.

TECHNICAL STUFF

Tachyons are a consequence of Einstein's theory of special relativity which, as you now know, brought together space and time. American physicist Gerald Feinberg came up with the term *tachyon* in the 1960s to describe a particle with imaginary mass that arises from a quantum field. Normal particles, ones that travel slower than the speed of light, increase their speed as energy is added to them. Tachyons increase their speed as energy is taken away. The lowest energy state of these particles is moving at the speed of light. To an observer in our fixed, normal-mass reference frame, these particles would appear to be moving backwards in time.

Although the existence of these particles is controversial at best, one theory suggests that pairs of virtual tachyon and anti-tachyon particles could be fluctuating in what's called the quantum vacuum. Results from this model seem to be consistent with current data on dark energy as well as the period of inflation that took place just after the Big Bang (flip to Chapter 13 for a quick review), and the results might also be promising for dark matter. However, this idea is still very much a preliminary theory.

Origins of dark energy 5: Questioning Einstein and gravity

One of the most recent thoughts on dark energy takes a hard look at Einstein's original theories of gravity and the space-time continuum. What if Einstein's theory of gravity isn't quite right? What if the treatment of gravity in general relativity needs an extra term, a field that causes the expansion of the universe to accelerate?

TECHNICAL STUFF

The current theory of gravity works well in addressing the observable universe; it governs the behavior of stars, planets, and galaxies. Any change to the theory of gravity, such as a modification or an extra term, would also need to support these relatively smaller objects. However, because dark energy might be visible mostly at the very largest scales, it's possible that observing galaxy clusters could lead to evidence of this different gravity behavior.

The standard of cosmology: The Lambda-Cold Dark Matter (ΛCDM) model

Ever heard the expression, "don't put the cart before the horse"? Just for a moment, let's locate your favorite Clydesdale in front of your electric-powered wagon so that we can put together the parts of a modern theory of cosmology. The Lambda-Cold Dark Matter (ΛCDM) model assumes three parts to the universe:

>> Baryonic matter (photons, electrons)

>> Cold dark matter

>> Dark energy

Now combine these pieces of the theory with some key observations built up over the 20th century, such as the expanding universe and the cosmic microwave background radiation. Bring more recent observations into the mix and you have what's considered the current consensus model of cosmology: ΛCDM. The most accepted variant of ΛCDM is called the standard, or 6-parameter, ΛCDM model, and it's basically a mathematical model that starts with the Big Bang and attempts to explain the universe from there. Not an easy feat by any standard!

The basic assumptions of ΛCDM start with general relativity and a further set of equations called the Friedmann-Lemaitre-Robertson-Walker equations. This theory assumes that the universe is made up of familiar particles, such as baryons, electrons, photons, neutrinos, and cold dark matter.

Remember that cold dark matter is dark matter with non-relativistic speeds that only interacts with regular matter through the force of gravity. The model also includes dark energy, in this model assumed to be expressed as a constant vacuum energy density, and it is described in the model as the cosmological constant Λ. The standard version of ΛCDM also makes an assumption that space is flat, suggesting that standard Euclidean geometry rules may apply.

ΛCDM tries to tie together observations of concepts such as CMB radiation, supernovae, galaxies, and quasars into a single unified model; this model is shown in Figure 15-6. Note here the changes in the expansion of the universe over time. Each of these types of observations is taken at a different redshift, providing a window into a different time period of the universe's past and allowing us to piece together a coherent story.

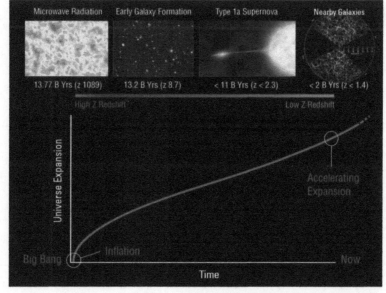

FIGURE 15-6:
Observations of the changing expansion of the universe over time, as part of the ΛCDM model.

Courtesy of NASA / LAMBDA Archive Team / Public Domain

Using ΛCDM with these assumptions, it's possible to arrive at a coherent model of the creation of particles and matter that

>> Starts with the Big Bang

>> Continues through inflation, the cosmic dark ages and cosmic dawn

>> Involves all the stages leading up to the formation of stars, galaxies, and the structures of the modern universe

New observations and new pieces of the theory continue to be developed; so far, ΛCDM seems to be our best attempt at a unified theory of the formation and evolution of the universe. We know it isn't the full story, but it is what we have today.

Now that we know where we came from, what about where we are going? It turns out that ΛCDM, dark energy, and dark matter are huge factors in determining what's called the critical density of the universe. This key parameter helps determine the ultimate fate of the universe — see Chapter 16 for more.

IN THIS CHAPTER

» **Dimming light and fading life as the sun makes its last hurrah**

» **Calculating critical density**

» **Freezing or ending in a fiery blaze?**

» **Heading out on the last train to nowhere**

» **Learning what might have come before**

Chapter **16**

The End of It All

They say that diamonds are forever (and by *they*, we are largely referring to diamond resellers). Diamonds are made of pure crystallized carbon that's arranged in a tetrahedral lattice. They're strong to be sure, but not strong enough to survive the end of times. Nothing is, and that's because all life and matter — including diamonds — are living on borrowed time.

Not to panic, though! You've got a good 5 billion years before it's time to start saying your last goodbyes. There was a time when scientists thought that the cosmos had been around forever and was essentially unchanging, but cosmology in the past 100 years has proven that the universe not only has a beginning point but also an ending.

You've now reached the concluding chapter of the detective story that is understanding the cosmos. All the evidence from cosmology theory has been building up to try and answer this question: What happens when the world (and the universe) ends?

No Refunds: What Happens When the Sun Explodes

"Mommy, what happened to the goldfish?"

Eternal questions of life and death are always difficult to answer, but they're a necessary part of figuring out one's place in the universe. Theoretical cosmologists spend their days trying to answer these questions on a cosmic scale. What are the main requirements for human life, and what will happen when those expire? How would you survive if the Sun became so hot that there was no way to cool down?

Running out of (hydrogen) gas

The short answer is that you wouldn't survive the end of the Sun, but let's back up a bit. Roughly 4.6 billion years ago, your best friends (dust and gas) met at a bar and hit it off. They danced the night away, spinning and spinning . . . until their combined internal gravitation caused a collapse and compression that became the Sun you know and love. As gravity pulled more and more matter into the Sun's core, pressure turned up the heat until the Sun reached its current core temperature of about 27 million °F (15 million °C).

REMEMBER

The thermonuclear fusion that converts the Sun's hydrogen to helium created enormous amounts of heat and light as a result. Thermonuclear fusion is the joining of lighter atomic nuclei into a heavy one, a process that releases significant heat and energy (see Chapter 5).

Much like your car's gas tank empties as you drive, or like your electric car's battery runs out of charge, the Sun will run out of fuel in about 5 billion years. When that happens, it won't be lights-out immediately, but the Sun will begin the same kind of end-of-life cycling that you learn about in Chapter 5. Without any hydrogen at its core to undergo fusion, gravity will compress the remaining helium at the Sun's core, though not enough to start up fusion with the helium as fuel at this point (the Sun would have to be a lot more massive to achieve that end goal).

TECHNICAL STUFF

This gravitational collapse of the Sun's core will cause it to heat up as the pressure increases. More heat and light will be released, and these processes cause the outer layers of the Sun to expand and grow into a red giant star. This kind of star is called a red star because even though it gives significant amounts of heat, it will be cooler than it is today. In part, this is because the Sun will get bloated out by the tremendous energy coming from its core, and when the energy gets spread out, it cools. The result is a red color coming from a much larger (and thus more luminous) surface.

Goodbye, life on Earth

The first thing that will happen, even before the end-of-life, red-giant phase of the Sun's evolution, is that its brightness will start to increase. It's estimated that this rate of increase will be a steady 10 percent in luminosity every billion years. The increased brightness means that the solar system's habitable zone, the distance from the Sun where liquid water will exist at the surface, will start moving outwards as well (see Chapter 7 for more on the habitable zone). For a relatively brief period, temperatures on Mars may be downright balmy. It's possible that there could be enough sunlight to make things very interesting on the icy moons of the outer solar system, especially Jupiter's moon, Europa. More on that later in this section.

This increase in the Sun's brightness won't be very good for life on Earth. Surface temperatures will increase and the oceans will eventually evaporate into steam, creating a thick atmosphere where this water vapor is broken down into hydrogen and oxygen. The Earth's atmosphere may then transition into a thick soup of nitrogen and carbon dioxide, similar to Venus's noxious atmosphere, as hydrogen gets lost to space and oxygen ends up bound up in surface rocks. If humanity has become a spacefaring race by this point, it would be far past time to escape out to Mars or the outer solar system, though even those havens won't be habitable for long.

As the dying Sun enters its red-giant phase, over the course of about 5 million years, it will begin expanding, eventually growing large enough that its outer layers could reach a radius of more than 100 million miles (170 million kilometers). Because the Earth orbits at a distance of 93 million miles (150 million kilometers), the inner planets Mercury, Venus, and maybe even Earth will be vaporized and absorbed by the expanding Sun. Maybe. It's possible that as the Sun losses mass, Earth will migrate outwards and save itself. Right now . . . it's predicted to be a close call with Earth staying just outside the Sun's surface. The red giant stage will last about a billion years. Figure 16-1 shows the end-of-life phases that the Sun is expected to go through.

As the expanding Sun reaches the orbits of Mercury and then Venus, our nearest planetary neighbors will become completely engulfed. At this point the red giant Sun would be so close to the Earth that temperatures would be unlivable for any human, animal, or plant matter; what's left of the oceans would boil off, and this is the point where our collective time on Earth would come to an end. Even if Earth manages to survive, not much will be left of our home planet but a burnt-out husk.

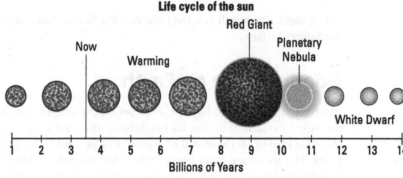

Life cycle of the sun

Now

Warming

Red Giant

Planetary
Nebula

White Dwarf

FIGURE 16-1:
End-of-life phases
of our Sun.

1 2 3 4 5 6 7 8 9 10 11 12 13 14

Billions of Years

SPENCER SUTTON / SCIENCE PHOTO LIBRARY

The Sun is expected to settle into a stable red giant phase that would last about a billion years. Even Mars and the outer solar system won't stay inhabitable during this time, so don't pack your bags for Jupiter just yet! During this time, the habitable zone could move out as far as today's orbit of Neptune or even Pluto. It's possible that some remnants of humanity could survive in the outer solar system. The real question becomes how much mass our Sun will lose, and how far out worlds like Jupiter (with its moons) will be able to migrate. Although it is likely barbecue time even for Jupiter, there is a possibility that life could find a way.

Although humanity's end may come at this point, it's certainly not the end of the story. After the Sun runs out of hydrogen in its outer core, gravity will compress our former flaming ball and it will collapse once more. The core pressure would finally be high enough to support about 2 billion years of fusion reactions in which the Sun would burn helium to make carbon and oxygen, but even this fuel will run out. At this point, the Sun's remnant core would collapse further and form a white dwarf. The outer layers of the red giant's remainder would be far enough away to escape the gravitational tug of the core's collapse, and these layers would keep expanding through space to form an object called a planetary nebula. The white dwarf, initially quite hot, would give off enough energy to light up the planetary nebula as a farewell signal for what's left of our solar system.

Omega Value of the Universe

Although our Sun is the center of our home system, it's not the center of the universe. The Sun's eventual death will affect everyone and everything on Earth, to be sure, but we need to take a step back and broaden our scope. The untimely demise of the Sun is unfortunate (for us) but fairly well understood in the grand scheme of things. The ultimate end of the universe, on the other hand, is the subject of much debate and many theories.

At the core of these theories is a concept called the *critical density parameter*. This term, known in the world of cosmology as Omega, is a measure of the average density of matter required for the universe to stop expanding. Omega can also be written as the Greek letter Ω.

REMEMBER

One of the revelations from Einstein's theory of general relativity (see Chapter 15 for the details) was the idea that gravity curves space around it. Implicit to this curvature is the density of that space. There are three ways to consider this density with respect to expansion; take a look at Table 16-1 for a summary.

TABLE 16-1

Expansion Parameters for the Universe

Density, Omega Value	Universe Shape	Expansion Type
Low, Omega < 1	Open	Universe keeps expanding; Big Freeze
High, Omega > 1	Closed	Expansion stops, universe collapses; Big Crunch
Critical Density, Omega = 1	Flat	Expansion stops, but only after infinite period of time

The main idea here is that if Omega is greater than 1, the universe is a closed system that will eventually collapse back into a point. If Omega is less than one, the universe is an open system that will keep expanding indefinitely. Finally, if Omega is equal to one, the critical density has been met. The result? You get a flat universe where the expansion will become asymptotically slower and slower, but will not stop until time reaches infinity. (This concept is common in math, where you have a value that gets closer and closer to a limit but never actually reaches it.)

REMEMBER

In addition to its implications for the future of the universe, Omega also becomes relevant in terms of the shape of the universe, as shown in Table 16-1. Remember that because matter can curve space and time, the critical density also controls the shape of the universe. Figure 16-2 shows three possibilities for the shape of the universe. One caveat is that these are two-dimensional representations of what is actually a very complicated three-dimensional topology.

If the density of the universe is high and the universe is closed, the geometry can be described as spherical; local light rays that are parallel at one point will eventually converge and touch each other. If the density is low, the universe is open and has a hyperbolic shape (like a saddle) in which parallel light rays eventually diverge away from each other. Finally, if the universe's density is at the critical density, its geometry can be described as flat, and parallel light rays will always remain parallel.

Spherical Space　　　　**Flat Space**　　　　**Hyperbolic Space**

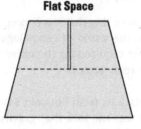
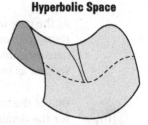

FIGURE 16-2:
Critical density
and the geometry
of the universe.

It's critical: The critical density parameter

So what actually is the critical density parameter? For our universe, astrophysicists have calculated that the critical density, ρ_c (ρ is the Greek letter rho) can be defined as

$$\rho\_c = (3H^2) / 8\pi G$$

where H is the Hubble parameter that describes the rate the universe is expanding at any given moment (the Hubble Constant, in Hubble's law from Chapter 9 is our current value), and G is the gravitational constant (covered in Chapter 2). For our universe, ρ_c works out to a critical density of about 9×10^{-27} kilograms per cubic meter at our current moment in time. This density is extremely low; for comparison, the concept here is similar to taking the matter in a drop of water and spreading it out over a volume the size of the Earth. This density works out to about one single hydrogen atom for every 4 cubic meters of volume.

Does it seem like this critical density is so small that our universe can't possibly be that empty? It's actually spot-on, and here's why: The universe is mostly empty space, and that empty space is really mostly empty.

TIP

The "mostly" here is key. Take a look at Figure 12-3 to see the percentages of the universe that're made up of regular matter (only 5 percent), dark matter (27 percent; this is material that only interacts with other matter through the force of gravity), and dark energy (the remaining 68 percent; this is some kind of weird energy density that could come from empty space itself.)

**TECHNICAL
STUFF**

You can dig deeper with a little math. Omega, written as the capital Greek letter Ω, is defined as the density of the universe in terms of a fraction of the critical density. It can be written as

$$\Omega = \frac{\rho}{\rho_c}$$

where ρ is the actual density of the universe, and ρ_c is the critical density of the universe. As shown in Table 16-1, if $\rho = \rho_c$, then the value of $\Omega = 1$ and the universe is flat. More on what that would mean in the sidebar "The Flat Universe."

Λ AND Ω ARE BOTH GREEK TO ME

Remember that Cosmological Constant, the one that Einstein added to help ensure that the universe was static and unchanging, and which he later referred to as his greatest mistake? That term, Λ, later made a reappearance as the vacuum energy density of space. So how does Λ relate to Ω? Yes, they are both Greek letters, but the connection runs deeper. After Λ was redefined as the vacuum energy density, or the energy density of empty space, it became closely associated with dark energy. Because one of the components of Ω is from dark energy, the complete description of Ω as the total density parameter includes a term called Ω_Λ or the effective density of the universe due to dark energy. In other words, Λ is part of total Ω.

TECHNICAL STUFF

Ω has to have contributions from regular matter but also from dark matter and dark energy. Ω is sometimes written as the sum of Ω_m (a factor produced by matter and the attractive gravity that it is subject to) plus Ω_Λ (Λ, or the Greek letter Lambda, is a factor produced by space itself and its intrinsic springiness and repulsive forces).

Here's another way to think of Omega:

$$\Omega = \Omega_m + \Omega_{rel} + \Omega_\Lambda$$

In this case, Ω is again the total density parameter. Ω_m is the total mass density, which includes normal mass made from baryons (stars, planets, interstellar gasses, and so on) plus dark matter. Ω_{rel} is the mass density of relativistic particles, those that travel close to or at the speed of light. These include photons and neutrinos. Finally, Ω_Λ is the effective mass density of dark energy; it's described by the resurrected cosmological constant, so this term is written to refer to Λ.

How do we calculate Omega?

Omega is critically (ha!) important, and there's more than one way of figuring out its approximate size. Two methods that have been used to try to make an initial estimate of the critical density of the universe are

>> The accounting method
>> The geometric method

In the accounting method, astrophysicists sum up the mass of all the objects within a certain volume of the universe in order to estimate the mass of that portion of the universe. It's fairly straightforward to estimate the mass of a galaxy or a galaxy cluster by looking at the motions of galaxies in a cluster, or by summing up the luminosity of visible galaxies (see Chapter 11 for a refresher here). This technique becomes more complicated when you add in dark matter, though, and it requires assumptions about the ratio of dark matter to luminous matter. The method, with its accompanying assumptions, allows you to estimate the total mass in a particular volume of space. The mass divided by the volume therefore provides an estimate of the density.

In the geometric approach, arguments about the shape of the universe are used as a thought experiment to determine whether Omega is less than or greater than one. These are the two main scenarios:

>> If the universe was open, you'd expect that because parallel lines diverge over time, the density of galaxies in the distant past would be observed to be greater than predicted.

>> Conversely, if the universe was closed, parallel lines could eventually converge over time; the implication is that the density observed for distant galaxies would be less than predicted.

Interestingly, both the accounting method and the geometric method currently provide a value for the density of the universe that is very close to the critical density, and this value makes Omega seem to be very close to a value of one. Do we actually live in a flat universe? Perhaps, but these are just initial estimates. Keep an eye out for future data.

WMAP and Planck missions

Let's not forget the role of our elusive yet ever-present friend, dark energy. Using the method in which we divide up the values of Omega into contributions from the mass density, the relativistic particles, and the dark energy, we can estimate these three components and add them together. Note that by convention, when these measurements are generated they refer to the values at the current time, so an extra subscript of 0 is often added.

To measure these values, NASA launched the Wilkinson Microwave Anisotropy Probe (WMAP) spacecraft, in operation from 2001 to 2010. The European Space Agency (ESA) launched a follow-up mission, the Planck spacecraft, and this mission operated between 2009 and 2013. WMAP and Planck's main science goal was to measure the detailed variations in the Cosmic Microwave Background radiation with unprecedented detail; Chapter 13 is your source for more information on the CMB.

These subtle variations ("anisotropy") in the CMB are thought to date back to acoustic waves in the structure of the universe, soon after the Big Bang. Because the CMB dates back to when the universe was only a few hundred thousand years old and we can survey it in all directions, it provides us a chance to completely study one moment early in the universe. Measurements of the subtle ripples in the CMB allow astrophysicists to measure Ω_m and Ω_Λ with unprecedented accuracy.

The values measured from WMAP and Planck give the following pieces of information:

>> Mass density $\Omega_{m,0}$ was estimated to be 0.27 +/- 0.04. Baryonic matter only gives you $\Omega_{m,0} = 0.44 +/- 0.004$, so it only makes up about 17 percent of the matter of the universe, leaving the rest to be classified as dark matter.

>> $\Omega_{rel,0}$ includes relativistic particles such as photons and neutrinos. Measurements from WMAP suggest that $\Omega_{rel,0} = 8.24 \times 10^{-5}$. The contribution from relativistic particles is therefore so much smaller than that from matter, that this term can mostly be neglected.

>> $\Omega_{\Lambda,0}$ is the fraction of the mass of the universe that is due to dark energy. This value was estimated by WMAP to be $\Omega_{\Lambda,0} = 0.73 +/- 0.04$. This means that dark energy is responsible for 73 percent of the effective mass of the universe.

These current values, when added together, arrive at

$$\Omega = \Omega_m + \Omega_{rel} + \Omega_\Lambda = 1.02 +/- 0.02.$$

This value closely approaches the critical density of $\Omega = 1$! Note, though, that these are the current-day values for these parameters. It looks as if in the very early universe, radiation dominated, then matter dominated, and finally, we are currently in an era where dark energy dominates.

Future work on critical density

Other measurements can also be added to the estimates from WMAP and Planck as we further refine our understanding of Ω. Figure 16-3 shows the values of Ω_m and Ω_Λ from the CMB measurements of WMAP and Planck as the diagonal zone labeled CMB. Measurements can be added from Type 1a supernovae, and these are sensitive to the rate of expansion of the universe; visualize these as concentric ovals labeled SNe. You can also add Baryon Acoustic Oscillations, a measure of the amount of galaxy clustering and an indicator of the mass density parameter, shown at the vertical zone labeled BAO.

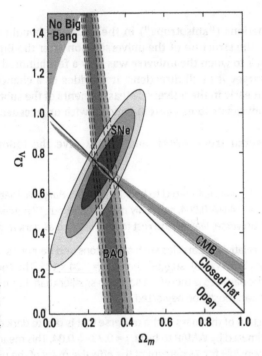

FIGURE 16-3:
Union plot of multiple methods of estimating the critical density of the universe.

TECHNICAL STUFF

All of these constraints seem to be consistent with the values suggested above; the mass density Ω_m is about 0.27 and the dark energy density Ω_Λ is about 0.73, meaning that the critical density is just about 1.0.

Ongoing research continues to work toward understanding the factors that might go into better estimates for Omega, namely dark energy and neutrinos. New telescopes like the James Webb Space Telescope are already receiving light from unusual, distant objects dating back closer to the origin of the universe. Newly discovered bright primordial galaxies dating back to only about 300 million years after the Big Bang might seem to contradict the ΛCDM model of cosmology, but could also indicate that other parameters in this model need to be tweaked. These parameters include the initial mass function, an assumption about the sizes of stars in a galaxy. More and more of these large, high-redshift galaxies are being found by JWST. These galaxies appear bright and also fully evolved, and contain major hallmarks such as accretion disks and central bulges.

JWST is also providing evidence for the expansion rate of the early universe, and there are indications that it might not be consistent with current estimates. One reason for this inconsistency could be the density of dark energy, and this density may be in flux rather than constant. This problem, known as Hubble tension, may also cause estimates of the age of the universe at a particular redshift to be revised. For example, those huge, well-formed galaxies that seem to date back to about 300 million years after the Big Bang might actually come later, more like 500 million years post–Big Bang. This revised dating would allow more time for stars and galaxies to grow and evolve, consistent with the ΛCDM model. (This fate was already experienced by one galaxy which had its age updated from 240 million years to 1.2 billion years after the Big Bang!)

Although observational evidence is coming in, theoretical models are also expanding our ideas for the ultimate fate of the universe, based on the values of Omega and its components.

The Big Freeze: An End of the Universe Theory

Here's yet another thought experiment: Would you want to know the exact time and date of your own death? Suppose that knowledge isn't power, and that you wouldn't be able to do anything to change or control this date. Would you still want to know? Those who say yes are likely either inherently curious or want to make sure they've accomplished everything on their bucket list while there's time; people who say no may not want the stress or worry to affect what time they have left.

On a cosmic scale there's less urgency because it's well established that our individual lifetimes may be short but our planet's got at least a few billion good years left. Still, the story of cosmology has to have a beginning and an end. Although the question of how the universe began is incredibly important to astrophysicists worldwide, there's an equally important question — how will the universe end?

There are three main theories that we explore here concerning possible universal endgames. Figure 16-4 illustrates them, and shows how dark energy plays a huge role in our ultimate fate.

The first is known by two apparently contradicting names, "heat death" and "the big freeze," and is listed as "indefinite expansion" in Figure 16-4. We dig in here and sort out the confusion.

The last stars burn out

In Table 16-1, we listed three possible categories of Omega values. In the first case, in which Omega is less than 1, the total density of the universe is less than the critical density. Or, more likely, the dark energy component will dominate and the expansion of the universe may even accelerate over time instead of slowing down.

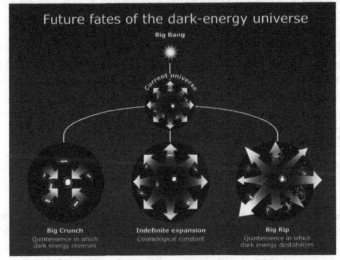

FIGURE 16-4: Three possible fates for the future of the universe, and the role of dark energy.

Courtesy of NASA/ STScI/ Ann Field / Public Domain

TIP

Under this scenario, the universe would expand and become so spread out that, eventually, the last stars burn out and no new stars could form. Even the last galaxies would spread out, ultimately resulting in a scenario where all that's left are individual atoms spread out through space. The kicker? These atoms would be too far apart from each other to interact in any way. Everything would be cold, static, and unchanging — it's a triumph of entropy.

TIP

Let's explore this cheery scenario a bit more. Our current universe is filled with stars, galaxies, black holes, and maybe even with life. Astronomers sometimes say that we are living in a luminiferous era, because it's a galaxy and universe filled with light. New stars are still being formed in our galaxy and in other galaxies, but the rate of maximum star formation is now behind us. The universe is about 13.8 billion years old, and star formation peaked about 10 billion years ago, only 3.8 billion years after the Big Bang, and it's been slowing down ever since.

In this scenario, you have a universe that would continue expanding forever, and that expansion would be gradually accelerating due to dark energy. The acceleration factor here would be constant, and the amount of dark energy would be related to the Cosmological Constant. In addition, the amount of mass in the universe would remain constant; as the universe expands, its matter would be spread around through more and more volume. The density of the matter in the universe therefore would get smaller and smaller over time as the universe gets more and more spread out.

Eventually, about a trillion years from now, the last stars would be formed. These would likely be low-mass, red dwarf stars. Star-forming regions simply wouldn't have the mass density to form anything larger; after this last round of stars form, there wouldn't be enough co-located mass for gravitational attraction to take place and fusion to ignite.

The last low-mass, red dwarf stars to form would have long lives, like most low-mass stars, but even they wouldn't be able to last forever. In about 100 trillion years, the last stars would burn out. The last light in the universe would then go out — forever.

Over the horizon: Galaxies beyond view

What would the sky look like during this period? Remember, most of the solar system would be long gone by this point because of the Sun's red giant phase, but imagine that humanity has found a new home on some rogue planet unaffected by any star's death. As the universe continues to expand and the rate of expansion continues to increase, more and more galaxies would be past the horizon over which we could see them. The reason for our future failure to experience these friends from afar would be that light from distant galaxies can only travel at the speed of light, a fixed speed, but the universe would be expanding at a faster rate.

The most distant galaxies would already be expanding away from us at a speed faster than the speed of light, so we would never be able to see their light. We could only see the past light that these galaxies would have emitted when they were closer to us and, therefore, not expanding away as quickly. This threshold would get closer and closer to us; when an increasing number of galaxies were over the visible horizon, we would start to see fewer and fewer galaxies in the sky.

TIP

As the final galaxies disappear into the distance, only the Local Group, our own galaxy cluster, would remain nearby because its components are gravitationally bound to each other. The Milky Way, Andromeda, and Triangulum galaxies, plus the variety of satellite dwarf galaxies (flip to Chapter 11 for the details), would eventually merge into one mega galaxy. This new galaxy would stand alone — nothing else would be visible in the sky, or in the universe.

Eventually, even the stars in our local mega galaxy would randomly encounter each other, disrupt each other's orbits, and move too close to the black hole at the center of our galaxy. Either the black hole would absorb these stars, or they would be ejected from the galaxy where they'd head out into empty space all alone. After about 10^{20} to 10^{30} years, there wouldn't be any large-scale structures left in the universe. All remaining stars would become isolated objects in a dark cosmic sea, utterly alone.

Such a degenerate era

TIP

But wait! If that cheery thought wasn't uplifting enough, our downward spiral would continue. When all the stars burn out, the universe would be in a degenerate era that would last for quintillions of years. (A quintillion is a billion billion.) Whatever stars still remain would become burnt-out husks that no longer generate any heat. There would still be neutron stars, white dwarfs, and black holes, but when they completely cooled none would emit radiation. White dwarfs would transition into a theoretical category called black dwarfs, and the universe would be cold and dark as it continues its path of expansion.

Further down the road, these remnants of stars and other objects would slowly break into individual atoms due to random quantum fluctuations. After about 10^{65} years, there would no longer be any large physical objects; those leftover burnt-out stars would be completely disintegrated and all that remains would be a vast sea of cold, diffuse atoms.

REMEMBER

Black holes would last the longest before fizzling out, but even black holes give off radiation, although only a little bit at a time. As we discuss in Chapter 8, black holes do give off Hawking radiation, but this process is extremely inefficient — a black hole might only give off one particle per year. However, the inexorable march of time can wait out everything, even black holes, and after 10^{100} years they'd be completely gone.

Remember those individual atoms? Even they would start to come apart. If protons are stable over these long-time scales — and that's a huge if, given that there's no way to test this in a laboratory — they may be the largest objects in the universe until they, too, decay after about 10^{200} years.

Heat death of the universe

REMEMBER

If dark energy continues to dominate and the rate of the expansion of the universe continues to accelerate, the universe would eventually reach a phase known lovingly as "heat death." The physics behind heat death can be summarized by the second law of thermodynamics (see Chapter 2 for a refresher): Over time, entropy tends to increase in an isolated system. Because entropy is a measure of the disorder of a system, in terms of the universe, an organized system of stars and planets and galaxies will tend to become more and more disorganized over time as entropy increases, until all you have is a sea of particles.

THE FLAT UNIVERSE

An interesting variant on the Big Freeze is the special case of Omega = 1. What happens if the universe is actually flat and the mass density is equal to the critical density (see Table 16-1)? This universe would be teetering right on the edge of expanding forever and contracting down to a point again. If $\Omega=0$, the universe would keep expanding, but the rate of expansion would get slower and slower until it approaches (but never quite reaches) zero. This version of the universe would be similar to the Big Freeze but because it involves less expansion towards the end, it might never actually reach full heat death.

TECHNICAL STUFF

Assuming that the laws of thermodynamics would continue to operate as we approach our collective demise, the entire universe would eventually reach complete thermal equilibrium. Specifically, no part of the universe would be any warmer or colder than any other part of the universe, and the universe would be in a state of maximum entropy where no more work can be extracted from this system. This end state is called the heat death of the universe. The temperature would be completely uniform but would continue to drop, approaching absolute zero but never quite reaching it. For this reason, this theory is also called the "Big Freeze" theory.

TECHNICAL STUFF

What would happen next? Because we've never seen the end of a universe, no one really knows. Perhaps our current universe is all that there is, or maybe a new form of physics will cause a new Big Bang which could launch a new universe. One weird possibility comes from the theory of quantum gravity. Random fluctuations at the quantum level suggest that even after the universe reaches this state of maximum entropy, previous states could stage a comeback. Under this theory even the solar system could be reassembled, given enough time — and one thing the heat death story has is time! It's even possible that the entire solar system and universe could return and then proceed towards heat death one more time. That's a slightly more cheery thought, isn't it?

The Big Rip: Another End of the Universe Theory

Infinite expansion and death by cosmic loneliness is not the only possible outcome for the end of the universe. Remember, dark energy, making up about 68 percent of the universe, could be responsible for an accelerating rate of cosmic expansion. Now suppose that if, instead of being constant, dark energy changes and evolves over time. That level of change could cause the rate of expansion to increase faster and faster, potentially disintegrating everything from molecules to stars to entire galaxies, long before reaching the heat death of the universe. Instead of the universe spreading into nothingness with the Big Freeze, what if it disintegrated from within? This end-of-life theory is known as the Big Rip.

Dynamics: Dark energy changing over time

In the Big Freeze theory, dark energy is a property of the cosmos that's related perhaps to the vacuum energy density of empty space. It causes the expansion of the universe to accelerate and is treated as a constant in the Big Freeze scenario. It's also thought to be related to the Cosmological Constant (see Chapter 15).

But what if, instead, dark energy was a dynamic property that could change over time? If dark energy was in fact an energy field, what if it grew more powerful with time? If this scenario came to pass, dark energy could potentially start stretching out the universe faster — but it could also build up within objects themselves.

REMEMBER

This scenario is related to the idea of quintessence from Chapter 15; quintessence is an undetectable energy field providing a repulsive force that increases the rate of the expansion of the universe.

Phantom dark energy

Dark energy that can change over time is called dynamic dark energy; it can also be called phantom dark energy. Rather than the amount of dark energy staying the same over time in one particular volume of space, dark energy could increase and build up over time in the dynamic model. Phantom dark energy could build up within a galaxy over time, for example, and its repulsive force could actually push the galaxy apart.

In this scenario, rather than creating more space as the universe expands, dark energy instead begins building up and pushing apart objects as time goes on. Starting at the largest scales, galaxy clusters would be the first to go. They'd be pushed apart by this repulsive force, and their constituent stars would break free of their gravitational bindings before drifting apart from each other. Stars would eventually explode, and the disruption of dynamic dark energy could move on to smaller and smaller scales.

TIP

Molecules and even atoms would eventually be torn apart, and the whole universe would be a disrupted mess of subatomic particles. Not dire enough? For its big finish, the universe itself would be completely torn apart and space-time itself would be disrupted. Unlike the Big Freeze, where the final steps of disintegration of the universe into subatomic particles stretch out over untold billions of years, in the Big Rip the final stages of disintegration get faster and faster, with star systems flying apart months before the Big Rip. Stars and planets themselves would be torn apart in the final minutes of the universe, and atoms would be pushed apart in the last few fractions of a second before the disappearance of space-time.

TECHNICAL STUFF

Although this scenario is definitely supper-table conversation fodder, cosmologists think it's a lot less likely than the Big Freeze concept. The key parameter in judging whether or not the Big Rip could happen is an equation of state parameter w, defined as the ratio between the pressure of dark energy and its energy density. If w is between -1 and 0, the expansion of the universe does accelerate, but dark energy doesn't build up over time and we are in the Big Freeze scenario. For a Big Rip to happen instead, w has to be less than -1, and its energy density actually increases as the universe expands.

Observations of galaxy cluster speeds measured by the Chandra X-ray Observatory have been used to estimate w, and have given values between -0.907 and -1.075. Per these measurements, the Big Rip can't be ruled out, at least not based on current observations. On the plus side, if this Big Rip does come to pass, it won't be for at least 150 to 200 billion years.

The Big Crunch: Yet Another End of the Universe Theory

For a third possible scenario on how it all ends, suppose the universe keeps creating, gathering matter, and expanding . . . until it doesn't. What if there were a maximum point of expansion, beyond which gravity would halt further expansion and compress the entire universe back down to a single point? This theory, that expansion is replaced with contraction until all matter is reduced to nothing, is known as the *Big Crunch*. In Table 16-1, this would be the scenario in which $\Omega > 1$.

Expanding, then shrinking

The idea of a Big Crunch as a kind of bookend to the Big Bang has been around for years. The basic idea is that the universe is currently expanding, but it would someday reach a maximum expansion and then start contracting again. In this scenario, the universe eventually re-collapses. The theory was popular after the initial discovery of the expansion of the universe but is now thought to be less likely due to the measured increase in the expansion of the universe from dark energy.

TECHNICAL STUFF

If the Big Crunch were to occur, the scenario would be high on drama. The galaxies in our night sky would all rush towards us, rather than away from us, and would eventually collide and merge. Instead of the redshift we are used to in electromagnetic radiation, light and other signals from faraway sources would become blueshifted and would result in higher frequency radiation.

REMEMBER

This higher-energy radiation would bring some serious consequences because even visible light would be pushed to ultraviolet, then x-rays, and gamma rays. This external radiation would actually cook stars from the outside in, causing thermonuclear reactions on the surfaces of stars that would lead to their ultimate destruction. This stellar destruction would first start with cooler M-type stars and then proceed to hotter and hotter stars, which would be heated externally to the point of disintegration.

Unlike the Big Freeze scenario with its slow, inexorable heat death, the Big Crunch would be hot and fast rather than slow and gentle. Hot O-type stars would be heated and boiled off about 100,000 years before the Big Crunch, and in the last few minutes atoms would break up and fall into giant merging black holes. The universe would end in a giant fireball with infinite temperature taking up zero space, and ultimately would destroy space-time completely.

Spoilers: Dark energy could get in the way

Does the idea of dark energy support a Big Crunch theory? Not really, and here's why. For the purpose of this thought exercise, consider *matter* to be all matter — dark and baryonic matter. A universe in which a Big Crunch could take place would be one dominated by matter, where the gravitational attraction of all this mass eventually overcomes the initial expansion, bringing everything back together to a point.

Dark energy could make this theory invalid because, with its intrinsic repulsive properties, it's thought to be causing the expansion of the universe to speed up. We don't currently have a way to turn around this expansion and produce a contraction instead, as would be required for a Big Crunch to take place. Some new theories suggest that a Big Crunch scenario could arise due to a dark energy fluctuation, but as yet there's no observational evidence supporting this theory.

Something Before Nothing: Did Anything Come before the Big Bang?

Take a balloon and blow it up. Now let the air out so that it deflates completely. Can you blow it back up again?

Think about how that scenario could apply to our universe. Is our universe the first and only to have ever existed, or did other universes come before ours? If the Big Freeze theory is correct and the universe were to keep spreading indefinitely, you'd expect that this infinite spread would be — well — infinite. The Big Crunch theory lets you take it a step further in the thought experiment. If our universe expanded and contracted, what came before the beginning and what happens after the end? If the universe contracts back down to a point, could that point become a singularity for a second Big Bang? What if ours wasn't the first Big Bang? And why does our universe, and the laws of physics, support life? We tackle that last question first.

Anthropic principle: Why do the laws of physics even allow matter and life?

There are certain choices the universe seems to have made to allow life to exist. Fundamental atoms facilitated the creation of stars and planets. Our Earth formed when our BFFs, dust and gas, united via gravity to create a terrestrial planet suitable for life. Over the past half billion years, that life evolved into the human forms we inhabit today, and the universe has granted us permission to study it — and for you to read this book.

Of course, the preceding paragraph attributes intention to what very well could have just been a series of random occurrences — but all the same, here we are in a universe that supports life, at least life as we know it, and we have evolved as a species sufficiently enough to invent science and telescopes and start questioning the cosmos. When you reach these big questions of intent and causality, the line between a science-based field like theoretical cosmology can seem to merge and blur with basic questions of philosophy, and even theology.

Philosophical cosmology is a type of cosmology that attacks the study of the universe's form and structure from a philosophical perspective. This field asks the big questions, such as

>> What would happen if things had begun differently?

>> Why do we live in a universe that seems so finely-tuned to supporting life?

A response to this second question is the *anthropic principle*, a term first coined by American astronomer Robert Dicke. The main idea of the anthropic principle is that observations of the universe can only be made in a universe that is capable of supporting life. This principle is also sometimes called the observation selection effect. If the laws of physics in the universe were such that atoms could not bind together into matter, or dark energy ripped apart stars and galaxies before they could really take shape, then there would be no way that life could exist. By this theory, the simple fact that we are here to observe the universe means, on some level, that the universe had to be fine-tuned to support life.

There are many versions of the anthropic principle, but in general they can be divided up into two categories: the weak and strong principles:

>> The **weak anthropic principle** relies on selection bias. Given a huge number of universes to choose from, life can only observe a universe in which life is possible. One answer to the weak anthropic principle is the idea of multiple universes. In Chapter 15, we talk about how string theory can lead to the idea of the multiverse, that there could be multiple parallel universes existing simultaneously. Cosmology leads you to another question — what came before this universe and what might come afterwards? If there were multiple iterations of the universe, is the instance we are living in just one of a very few, among perhaps millions of versions, that could support life as we know it?

>> The **strong anthropic principle** goes a step further into a universe which is in some way compelled to evolve to support life. This idea skirts the line into theology or religion, implying that there could be an intentionality about the universe and the way that the laws of physics work together. Or it can arise from a principle of quantum mechanics, in which the universe must be observed in order to exist (and therefore requires an observer, like us).

Before and after: The cyclic universe theory

Multiple parallel universes existing at the same time are one possibility. An end-of-days scenario like the Big Crunch or the Big Rip leads you to think about what could come next — or what could have come before.

The cyclic theory poses an idea about what might have come before the Big Bang. Or, shall we say, "our" Big Bang. The cyclic model of the universe suggests that the universe has undergone repeated cycles of expanding and heating, contracting and cooling. Perhaps there have been multiple or even infinite Big Bangs; the universe starts from a point, runs out its course, and then at some point is finished with that iteration and begins anew.

Scientists on Earth don't have any way of measuring what came before the Big Bang, but just suppose the universe began as a point before expanding and contracting back down to a point — almost like an inverse Big Bang, also known as the Big Crunch theory. A scenario where the universe contracts down to a point mirrors the Big Bang but in reverse, so it's easy to imagine such a scenario recurring over and over again. A scenario like the Big Freeze is harder to end, but given enough time, those random quantum fluctuations may also work in a recurring cycle.

TECHNICAL STUFF

Could the universe enact another Big Bang, and could there have been other Big Bangs before ours? Was our universe even the first? There's no particular reason to think it was. Each new version of the universe, after each new Big Bang, could start off with a slightly different set of initial conditions; these conditions would involve different distributions of random fluctuations. Such slight differences could lead to huge changes in the distribution of matter and energy, or even in the laws of physics themselves. Some of these universes, perhaps many or almost all of these, could have been completely incompatible with life. We do, however, have a single data point that says that life was, and is, possible. Thank you, universe, and a big thanks to the anthropic principle!

Now That We're at the End — How Will It End?

You now have gotten a birds-eye view of the current scientific theories about the end of the universe. As noted, though, these are all theories — suppositions, propositions, and any other -*itions* that come to mind about principles and ideas supported by research, but not statements of fact. How do scientists know what's going to happen when the universal lights finally go out?

Science is built on the idea of the testable hypothesis. Scientists, astrophysicists included, are always thinking up new ways of testing out their latest brilliant theories. Clever experiments and new, state-of-the-art detectors and telescopes can often yield insights, confirming or disproving popular theories. These tools can also produce completely unexpected results that require a new theory.

Everything we've learned about the universe allows us to study not only where we came from, but also where we're going. We're at the cusp of an era where humans not only can observe but also understand. Who knows where future observations will lead! At least we have time, given that the Sun will be around for another 5 billion years, and the universe longer still. Here are some areas of cutting-edge cosmological research.

Getting Higgy with it: Vacuum decay and the Higgs boson and field

Although Heat Death, the Big Crunch, and the Big Rip all have their proponents, some scientists lean towards the theory of vacuum decay as a means to end it all. Consider stability in the universe — and no, we don't mean having tires with good traction, or keeping a pair of hiking poles in your trunk.

TECHNICAL STUFF

The moments after the Big Bang were filled with instability and phase change — the transition when any substance changes rapidly, as in from a solid to liquid or liquid to gas. After the fundamental forces of the universe (electromagnetism, gravity, the strong force and the weak force) phased away from the initial unified force resulting from the Big Bang, a period of general stability prevailed in the universe.

TIP

The Higgs boson is a tiny subatomic particle, predicted by the standard model of particle physics, that was recently observed experimentally for the first time at the Large Hadron Collider (LHC). The LHC is a high-energy particle accelerator where particles are, as the name suggests, accelerated into each other at very high speeds so that they turn into energy — energy that can then go back to being matter thanks to $E=mc^2$. If the energies are just right, you might get a tiny particle that is sometimes called the *God Particle*. It's related to the Higgs field, which, in the standard model of particle physics, is an invisible energy field giving mass to other particles. This energy field is otherwise invisible, making the detection of the Higgs boson a very big deal indeed.

The recent detection of the Higgs boson included a measurement of its mass, about 126 billion electron volts (or 126 times the mass of a proton). This mass is sufficient to keep the universe just on the side of stability through the Higgs field. However, that stability may be living on borrowed time.

Right now, the energy of the Higgs field is in a state of minimum potential energy. However, the field also has another stable energy state. This one would lead to a fundamental change in the laws of physics as we know them, and ultimately the destruction of the universe.

Fortunately, this second stable energy state would require a huge, implausible input of energy to reach. Unfortunately for us, however, particle physics also allows for an effect called quantum tunneling. Per this condition, the Higgs field could skip over the hard parts of climbing that energy mountain, and suddenly transition into a different, dangerous energy state.

Quantum tunneling could cause an entire cascade of effects that would change the laws of physics so that they no longer support physics (and life) as we know it.

This event could trigger a "bubble of vacuum" spreading and engulfing the universe at the speed of light, such that there would be no warning and no way to detect it.

The process undertaken here is called vacuum decay. Under this theory the universe would collapse into a black hole, or the quantum event bubble could spread until the entire universe is destroyed and ceases to exist. The detected mass of the Higgs boson suggests that the universe is on the brink of stability, so even a relatively small change in the Higgs field could tip us over into destruction. Future observations that might either confirm or deny this hypothesis may come from better observations of dark matter and how it interacts with particles on a subatomic level.

TECHNICAL STUFF

Another possible solution to the problem of vacuum decay could come from supersymmetry, a part of the standard model of particle physics stating that each particle not only has its own anti-particle but also its own supersymmetric particle. If these supersymmetric particles are found, they could help stabilize the universe and prevent unexpected obliteration via vacuum decay. So far, all tests of supersymmetry have failed, but astrophysicists keep looking.

Future observations of cosmic microwave background radiation and dark energy

Observations of the subtle variations in the cosmic microwave background radiation from the COBE, WMAP, and Planck spacecraft have helped build up a better understanding of such fundamental cosmological quantities as Ω_m and Ω_Λ. Although these space missions have ended operations, ground-based observations are continuing to refine our understanding of the CMB.

TECHNICAL STUFF

One type of observation uses telescopes to observe faint curling in the ripples of the CMB due to gravitational waves. These patterns are called B-mode polarization, and are being observed with a telescope called BICEP3 located in the cold dark of the South Pole. The results, when combined with data from WMAP and Planck, help shed light on the inflationary era that took place soon after the Big Bang. Stay tuned; more observations are yet to come.

Other research seeks to explain the so-called Cold Spot in the CMB, a large region with a lower-than-average temperature. A consortium of scientists used a very sensitive camera on a telescope located high in the Chilean Andes to create a database of more than 300 million galaxies. Data from this project, called the Dark Energy Survey, was used to create a map of dark matter, including in the direction of this CMB cold spot. This data was critical in finding that the region is a void with far fewer galaxies than expected. The presence of similar voids could have implications for the acceleration of the expansion of the universe, and could help provide evidence either for or against the ΛCDM model. This area of research is also ongoing as more observations take place.

Clues from JWST and the earliest galaxies

Finally, the new James Webb Space Telescope is starting to reveal the earliest history of the universe. Observations of the earliest galaxies might give us info by aiding our understanding of the world's beginning, making us better able to understand its end.

REMEMBER

In Chapter 14 you learn that the JWST is just beginning to find galaxies at the highest redshifts yet; these are the earliest times, back only about 300 million years after the Big Bang. Observations of these early days may show some of the first, no-metal Population III stars, and the structure and distribution of these early stars and galaxies will be important tests of our theories.

All these observations are ongoing as we humans use the best of our Earth-based and space-based observation powers to better understand the universe. Whether we are here through random chance in the one universe among billions that had the conditions to support life, or whether life itself is a cosmic certainty in a universe designed to support it, the study of astrophysics will continue investigating some of the biggest questions humanity can ask: Where did we come from, where are we going, and how will we get there?

We hope you've enjoyed this whirlwind journey through the world of astrophysics, and that you keep exploring the universe!

5

The Part of Tens

Discover ten scientists who paved the way for our understanding of astrophysics.

Explore ten critical space missions that have revolutionized our view of the universe. Learn about the Hubble Space Telescope, the James Webb Space Telescope, and everything in between.

Chapter **17**

Ten Scientists Who Paved the Way for Astrophysics

The field of astrophysics has been building for thousands of years. It began with early Mesopotamian records around 1000 BCE and spans through ancient Greece, Rome, and China, as well as into the Mayan empire and the Islamic world. From the Greek astrolabe to the Hubble Space Telescope today, key players have instigated research and innovation to keep building upon past knowledge, as well as creating a foundation for future astrophysicists. Here are ten of the most influential astrophysicists of the modern period.

Albert Einstein: 1879–1955

Raised in Munich, Germany and continuing his education in Switzerland, Albert Einstein famously worked in the Swiss Patent Office before becoming a Nobel-prize-winning professor of theoretical physics. He worked in Germany before immigrating to America in 1933, where he became a theoretical physics professor at Princeton University. Einstein's contributions to physics include revealing the curvature of space-time and creating theories of general and special relativity. His work on gravitational waves and black holes opened the world's eyes to the reality of how the universe worked, and paved the way for the entire field of modern cosmology.

Edwin Hubble: 1889–1953

American astronomer Edwin Hubble was leading research at the Mount Wilson Observatory (near Los Angeles, California) around the time that Einstein's theories of relativity were published. One of Hubble's first major discoveries was that the object formerly known as the Andromeda Nebula was actually a galaxy far beyond our own Milky Way. This discovery was key because it led to the realization that our own galaxy was only one of millions more. He was one of the first scientists to determine a relationship between galaxy distance and redshift, and this insight led to the conclusion that the universe was constantly expanding — a foundational principle of cosmological theory today.

Cecelia Payne-Gaposchkin: 1900–1979

Cecelia Payne-Gaposchkin was born and raised in England, and pursued her primary education there. In the 1920s, her options as a woman in science were limited and she moved to America under a Harvard fellowship. She spent her academic career at Harvard; since women were initially not allowed to serve as professors there, she had a series of research appointments until she became a full professor in 1956. Payne-Gaposchkin's primary contribution to astrophysics lay in her discovery that the main gas components of stars were hydrogen and helium. This conclusion was contrary to the prevailing ideas of the day; previously, it was believed that the composition of stars was similar to that of Earth, and this discovery was a game-changer in understanding how stars and galaxies formed. Additionally, her later studies of variable stars helped chart the course of stellar life cycles.

Karl Jansky: 1905–1950

American engineer Karl Jansky joins this list because of his immense contribution to understanding our galaxy. Jansky worked at Bell Telephone Laboratories as an engineer who invented the first radio telescope. Using one of the first antennae he designed, Jansky found a strange signal that he eventually determined to be coming from the center of the Milky Way Galaxy. Jansky is credited with inventing the field of radio astronomy; his findings were pivotal to the development of radio telescopes that would eventually capture significant data about galaxies and black holes.

Subrahmanyan Chandrasekhar: 1910–1995

Indian-American Subrahmanyan Chandrasekhar was a Nobel-prize-winning theoretical physicist who worked in the areas of star formation, composition, and life cycle. Educated in India and England, Chandrasekhar spent most of his career as a professor at the University of Chicago. His work highlighted how stars convert their hydrogen to helium over the course of their lives, ultimately collapsing under their own gravity. When a star of a specific size undergoes this transition, it collapses into a tiny, very dense star called a white dwarf. One of Chandrasekhar's biggest contributions to astronomy was determining that specific size; the *Chandrasekhar limit* states that a star with more mass than 1.44 times the Sun bypasses the white dwarf phase, explodes via supernova and turns into a neutron star. Put another way, the Chandrasekhar limit tells us how large a white dwarf star can be before it's no longer a white dwarf and is instead a detonating bomb!

Vera Rubin: 1928–2016

American astronomer Vera Rubin, like Payne-Gaposchkin, was a female scientist who was a bit ahead of her time. Rubin tried to study astronomy at Princeton but was turned down because women were not admitted at the time; fortunately for the world, she completed the master's program at Cornell and studied physics under the legendary Richard Feynman before receiving a PhD from Georgetown. She studied galaxy rotation throughout her career and uncovered myriad strange systems including a merging galaxy with two populations of counter-orbiting stars. Her work led directly to the discovery of dark matter, the substance scientists now believe comprises more than 80 percent of the matter in the universe. Rubin won the National Medal of Science award in 1993 for her work.

Kip Thorne: 1940–

Another Nobel prize winner in our midst! Kip Thorne is an American Nobel-prize-winning physicist who was heavily influenced by Einstein's relativity theory. After receiving a PhD from Princeton, Thorne became a professor at the California Institute of Technology. General relativity predicted the possibility of ripples, or waves, in space-time and Thorne gravitated (yes, we really said that) early in his career to the study of these gravitational waves. Among his most notable contributions to astrophysics was his foundational work on the gravitational waves that LIGO (the Laser Interferometer Gravitational Wave Observatory) would detect 100 years after they were first predicted by Einstein.

Stephen Hawking: 1942–2018

British theoretical physicist Stephen Hawking, author of multiple books including *A Brief History of Time,* is credited with making astrophysics, cosmology, and quantum theory accessible to us regular folks in addition to scientists and theoreticians. Hawking received his PhD from the University of Cambridge in the U.K., where he later returned as a professor. He began his career studying cosmology and physics, working heavily on the Big Bang theory of universe creation. Hawking is also well known for his discovery that a type of radiation, called *Hawking radiation,* could actually escape a black hole. His 2010 book *The Grand Design* not only refuted the idea that a deity could have created the universe, but presented his theories in such a way as to popularize astronomy, physics, and astrophysics. One of the most powerful thinkers of the 21st century, Hawking had the rare gift of making science interesting and applicable to everyone.

Jocelyn Bell Burnell: 1943–

British astrophysicist Jocelyn Bell Burnell began focusing on radio telescopes early in her career. Originally from Northern Ireland, Bell Burnell studied initially at the University of Glasgow, in Scotland. After receiving her PhD from the University of Cambridge, she worked at institutions in the U.K. and abroad, including Open University. Bell Burnell focused much of her research attention on quasars, also called quasi-stellar radio sources. These are celestial objects such as black holes that emit energy in the form of radio waves. In the process of studying them, Bell Burnell encountered regular pulses. She traced their patterns and located their source as neutron stars. This discovery of a new classification of astronomical object, later dubbed *pulsars*, was one of her most well-known contributions to astrophysics.

Alan Guth: 1947–

American physicist Alan Guth, professor at the Massachusetts Institute of Technology, has focused most of his research career on particle physics surrounding the creation and history of the universe. He's best known (so far!) for his theory of cosmic inflation, a way of explaining the rapid growth of the universe in the seconds following the Big Bang. His 1980 proposal of cosmic inflation described the role of gravity as a repulsive force within the false vacuum of the energy-dominated early universe (don't worry, gravity can't suddenly turn repulsive in other circumstances). This idea was revolutionary because previous assumptions about gravity as the force only of attraction, beliefs that had been in place since the time of Isaac Newton, were being challenged with this idea that gravity could also repel in this special moment less than a second after the formation of the universe. Guth's theory of repulsive gravity was a consequence of Einstein's General Relativity, and was confirmed by the 2014 detection of primordial gravitational waves.

Chapter **18**

Ten Important Space Missions for Astrophysics

Gathering information for a research project? You've got a few options: Go to the library and read up on your topic, conduct a few interviews, or make observations along with careful notes. Now suppose your research project is figuring out how the universe was created. Not too many people to interview! You can, though, collect data and make observations, and that's where spacecraft missions come into play. Earth-based telescopes provide an irreplaceable wealth of information, but some observations just can't be made from the ground. Here's a list of ten space missions that have given scientists the gift of learning about the cosmos.

Hubble Space Telescope (1990–present)

The Hubble Space Telescope (abbreviated HST) is named after none other than Edwin Hubble (1889–1953), the astronomer whose theory of universal expansion forever changed the face of cosmology. The HST was first proposed in the 1940s and finally launched in 1990 on the Space Shuttle Discovery. Its 8.2-foot (2.5-meter) diameter mirror and five different science instruments were designed to provide information in infrared, UV, and visible light wavelengths, and the telescope has captured thousands of objects for scientific study. See Figure 3-1 for an image of the Hubble Space Telescope.

There were early mishaps — a flaw in Hubble's mirror was preventing focus, and resulted in blurry images — but fortunately, HST was designed to be serviced by astronauts on the Space Shuttle. It took three years to design and build a module that would compensate for the slightly misshapen mirror, and in 1993 astronauts spent 35 hours repairing and upgrading the telescope in a series of spacewalks. Since then, the HST has returned immensely beautiful and telling images of black holes, galaxies, nebulae, and other deep-space bodies. A total of five servicing missions took place to keep the telescope operating at peak efficiency, including installing upgraded instruments and repairing broken ones. The final servicing mission in 2009 was one of the last flights of the Space Shuttle, but HST is still operating today with its final set of instrumentation.

James Webb Space Telescope (2021–present)

The James Webb Space Telescope (JWST) was named after James Webb (1906–1992), the head of NASA during the Apollo era. The JWST was launched in 2021 via an Ariane 5 launch vehicle. It uses a 21.3-foot (6.5-meter) mirror with near-infrared and mid-infrared instruments provided by the National Aeronautics and Space Administration (NASA), the European Space Agency (ESA), and the Canadian Space Agency (CSA). It's capable of taking the most distant images from Earth ever seen, but has also been used to capture planetary details that were previously impossible to detect.

JWST's infrared observations are made possible by a super-chilled detector, kept cold by a sunshield about the size of a tennis court that prevents direct sunlight from reaching most of the telescope. Figure 18-1 shows an artist's conception of JWST with its sunshield deployed. The shield results in a telescope temperature of about 50 Kelvin (−370°F, or −223°C). Rather than orbiting Earth directly, JWST orbits the Sun in a stable position beyond Earth's orbit called the L2 Lagrange point. This position allows the telescope to stay in line with Earth's orbit around the Sun such that the Earth, Moon, and Sun are always in the same direction to the telescope and can be shielded out simultaneously.

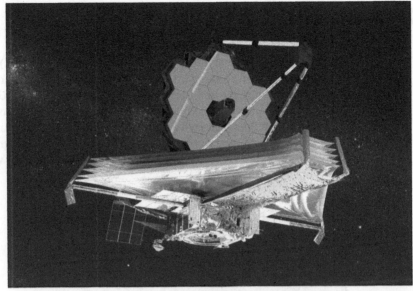

Courtesy of NASA GSFC / CIL / Adriana Manrique Gutierrez / Public Domain

FIGURE 18-1:
Artist's
conception of
the JWST
telescope with
the sunshield
deployed.

Kepler and TESS (2009–2018 and 2018–present)

Kepler, named after astronomer Johannes Kepler (1571–1630), was a NASA-led mission designed to look for transiting exoplanets. Launched in 2009 with a four-year "prime mission" period, Kepler targeted a small portion of the sky using a digital camera array, called the Kepler Photometer, to provide consistent, stable observations of its target region. Kepler was hugely successful, discovering more than 2600 planets orbiting stars beyond our own solar system and revolutionizing the field of exoplanetology.

Following Kepler was the Transiting Exoplanet Survey Satellite (TESS) mission. TESS was launched in 2018 from Cape Canaveral via a Falcon 9 rocket from SpaceX (and represented the first astrophysics mission to be coordinated between SpaceX and NASA!) TESS orbits in what's called a HEO, or high Earth orbit, and uses techniques similar to Kepler but is able to see 400 times more of the sky at once. So far, TESS has found more than 5000 potential exoplanet candidates, and the team is working to confirm which ones are genuine exoplanets using ground- and space-based follow-up observations — including follow-up with JWST!

SOFIA (2010–2022)

The Stratospheric Observatory for Infrared Astronomy (SOFIA) was an 8.9-foot (2.7-meter) airborne telescope that began regular flights in 2010. It was designed and built by NASA and the German Aerospace Center (DLR). Okay, not really a space telescope, but close! SOFIA consisted of a modified Boeing 747SP airplane outfitted with a door that could open in flight to allow the mounted telescope to observe, as well as instrumentation including special spectrographic and imaging equipment. See Figure 18-2 for an image of SOFIA in flight with its telescope hatch open.

FIGURE 18-2: SOFIA in flight with its telescope hatch door open.

Courtesy of NASA / Jim Ross

SOFIA flew much higher than a commercial airline up in the stratosphere. At altitudes of 38,000 to 45,000 feet, it was above more than 99 percent of Earth's atmosphere that blocks infrared radiation from observation. The observatory was able to gather data about solar system formation and key bodies within the solar system such as comets, asteroids, and planets, as well as more distant targets such as nebulae and galaxies. One of SOFIA's accomplishments was discovering water on the Moon's surface. The airborne nature of the telescope made it portable, and it could be deployed to study localized events such as transits and comets from both hemispheres. SOFIA's last flight was in 2022.

Chandra X-Ray Observatory (1999–present)

The Chandra X-Ray Observatory, named after Indian-American astronomer Subrahmanyan Chandrasekhar (1910–1995), was launched into space in 1999 with a boost from the Space Shuttle Columbia, and is still operating today. Its mission goals included using x-ray detectors to help answer questions about the formation and properties of the universe. This mission was able to locate hotter areas of space, such as those associated with galaxy clusters, and famously pinpointed supermassive black holes in many other galaxies.

Chandra is in an elliptical orbit that carries it very high above Earth, about 200 times higher than that of Hubble and about a third of the way to the Moon. This orbit allows it to spend roughly 85 percent of its time outside of the Van Allen radiation belts, a zone of charged particles surrounding the Earth that would interfere with Chandra's observations. Because x-rays would be absorbed by the typical convex mirrors used in optical telescopes, Chandra instead has four pairs of barrel-shaped mirrors that are impacted by x-rays at very slight angles. The x-rays are then focused onto four instruments that include a high-resolution camera, an imaging spectrometer, and two transmission grating spectrometers. These instruments allow incoming x-rays to be collected and characterized.

Spitzer Space Telescope (2003–2020)

Launched in 2003 on a Delta II Heavy rocket, the Spitzer Space Telescope was intended to send home images for 2.5 years but ended up lasting until its batteries wore down too far for safe maneuvering in 2020. The telescope was named after astronomer Lyman Spitzer (1914–1997), one of the first scientists to propose the idea of space telescopes. Spitzer was an ultra-sensitive infrared telescope with a 34-inch (85-centimeter) mirror. Spitzer's instrument payload included an infrared camera, an infrared spectrograph, and a multiband imaging photometer.

Like its successor mission, JWST, Spitzer had to be cool to see into the infrared. Without a sunshield, this required a cryocooler to chill the detector down to 5 K (−450°F or −268°C) using liquid helium. The telescope was placed in an Earth-trailing orbit around the Sun to keep Earth's thermal emissions from interfering with Spitzer's observations. When the mission's liquid helium supply ran out in 2009, the telescope was still able to operate in a limited warmer mode for another 11 years. One of Spitzer's most important discoveries lay in finding regions with newly formed stars, thereby providing scientists with more information about star and galaxy formation. Spitzer was also the first telescope to detect light directly from an exoplanet, and it even detected a new ring of Saturn.

Compton Gamma-Ray Observatory (1991–2000)

Deorbited in 2000, the Compton Gamma-Ray Observatory was launched in 1991 into low-Earth orbit from the Space Shuttle Atlantis. Figure 18-3 shows an image taken of it floating away from the Space Shuttle soon after deployment. The observatory was named after Arthur Compton (1892–1962), an American physicist whose work on gamma-ray physics was honored with a Nobel Prize. The mission included four instruments (EGRET, COMPTEL, OSSE, and BATSE) to measure gamma rays and x-rays. Unlike Chandra, Compton was intentionally placed into a low-Earth orbit so that it was underneath the Van Allen radiation belts where it could successfully take observations at high-energy wavelengths with less interference. This low orbit produced more drag on the spacecraft, however, and it required two orbital boosts via its on-board propellant.

FIGURE 18-3: The Compton Gamma Ray Observatory floating away from the Space Shuttle Atlantis soon after deployment.

Courtesy of NASA / Ken Cameron

Compton was able to locate entirely new types of galaxies with what we know today are supermassive black holes at their center and, in the process, provide visual documentation of gamma-ray bursts. It also discovered new blazar active galactic nuclei, and mapped the Milky Way at gamma-ray wavelengths. After one of the space telescope's gyroscopes failed in 1999, NASA planned for a controlled reentry over the ocean to avoid the potential for an uncontrolled crash if another gyroscope were to fail. The remains of the telescope safely crashed into the Pacific Ocean in 2000.

Fermi Gamma-Ray Space Telescope (2008–present)

The Fermi Space Telescope was launched in 2008 through a collaboration between NASA, the U.S. Department of Energy, and several different countries (Sweden, Italy, Japan, Germany, and France). The telescope is named after a pioneer in high-energy physics, Italian-American Enrico Fermi (1901-1954). It was launched on a Delta II rocket, and is still operating today in low-Earth orbit. Fermi is a follow-up mission to the successful Compton mission.

Fermi's science goal is to study the high-energy universe, specifically gamma rays, and to probe the extreme environments that generate cosmic rays. One of its key instruments is the Large Area Telescope, an imaging gamma-ray detector that allows it to perform an all-sky survey of gamma-ray sources. Another instrument on Fermi, the Gamma-ray Burst Monitor, can detect gamma-ray bursts, cosmic marvels thought to be caused by exploding stars, merging neutron stars, and other high energy events. Of note, Fermi has discovered and measured emissions from over 5000 gamma-ray sources to date. It can also study dark matter, and even search for evaporating micro black holes.

Herschel Space Observatory (2009–2013)

Launched in 2009 aboard an Ariane 5 rocket, the Herschel Space Observatory was built by the European Space Agency in partnership with NASA. Named after British physicist William Hershel (1738–1822), the scientist who discovered the infrared portion of the spectrum, Herschel was designed to detect far-infrared and sub-millimeter radiation wavelengths. It had a 11.5-foot (3.5-meter) primary mirror and three separate imaging spectrometers which provided high spatial and spectral resolution. Herschel's long-wavelength detectors required cryocooling, like other infrared space telescopes. To keep it cold and away from Earth's IR radiation, the spacecraft was placed in solar orbit at the stable L2 location, similar to

JWST. In addition to passive cooling the spacecraft also used a cryocooler; its liquid helium ran out in 2013, a year later than originally expected. The spacecraft was turned off at that time.

Herschel was designed to observe light from tiny particles of cosmic dust along with very cold objects that emit little heat. Some of Herschel's achievements include studying star-formation regions and locating water in the Milky Way; the answers may help fill in the blanks on water sources within our solar system. Herschel was also able to observe galaxies from a time when the universe was less than a billion years old, and during the peak star formation era when the universe was roughly 3 billion years old.

Nancy Grace Roman Space Telescope (planned 2027 launch)

The Nancy Grace Roman Space Telescope, named after NASA's first chief of astronomy, Nancy Grace Roman (1925–2018), is set to launch around 2027. Roman will be a wide-field infrared survey telescope, and will contain a 7.9-foot (2.4-meter) mirror similar to that of the Hubble Space Telescope. Its field of view will be about 100 times larger than Hubble's infrared instrument, and Roman will have both a Coronagraph and a Wide Field instrument. Like other IR telescopes, Roman will orbit the Sun at the stable L2 Lagrange point.

Roman was initially designed primarily to study dark matter and dark energy by surveying the distribution of galaxies in the sky. It is also expected to study exoplanets using a microlensing technique (see Chapter 7) that should allow the Wide Field instrument to detect thousands of exoplanets in the inner Milky Way Galaxy. In addition, studies show that the added Roman Coronagraph should allow the detection of objects more than a billion times fainter than their parent star, thus opening the door to far more precise research on exoplanets.

Glossary

accretion disk: A flat disk of plasma, gas, and other particles that surrounds a newly forming star or planet; created as matter spirals to the body via gravitational attraction and accretes into a disk.

active galactic nuclei (AGN): Extremely bright region at the center of some galaxies containing actively feeding supermassive black holes.

adaptive optics: A computer-driven technique for changing the shape of the telescope's mirror to make corrections for atmospheric distortion.

antimatter: Oppositely charged traditional matter; includes antineutrons, antiprotons, positrons.

atomic nucleus: The dense core of an atom containing protons and neutrons.

axions: Hypothetical, low-mass subatomic particles needed to explain some nuclear strong force interactions.

barycenter: The center of mass of a two-body system, such as a binary star or even a binary planet.

baryon: Subatomic particles consisting of three quarks; examples include neutrons and protons.

Big Bang: The way in which scientists explain the universe as forming from a point and then rapidly expanding.

binary star: A double star system in which the stars are bound to each other via gravity, and orbit their shared center of mass or barycenter.

black hole: A region in space having strong enough internal gravity that matter and radiation, including light, cannot escape.

blazar: An active galactic nucleus with jets oriented toward Earth.

blueshift: A way of describing the change in a light's wavelength as its source moves closer to the observer; as the light waves are compressed in the direction of motion, they move toward the blue side of the electromagnetic spectrum.

centrifugal force: A force acting on a body orbiting in a circular path that's pushing away from the center of that path.

centripetal force: A force acting on a body orbiting in a circular path that's pulling in toward the center of that path.

constellation: A grouping of stars that forms a pattern in the sky; most are named after ancient mythological figures.

cosmic ray: High-energy particles traveling through space nearly at the speed of light.

cosmology: The branch of astronomy that studies both the development and origins of the universe and its ultimate fate.

dark energy: A hypothetical force that opposes gravity, thereby encouraging the rate of expansion of the universe to accelerate.

dark matter: Matter that doesn't emit, reflect, or absorb light; observed indirectly through its gravitational effect on neighboring material, it cannot be observed directly.

Drake equation: A way of approximating the number of civilizations with intelligent life within the Milky Way Galaxy; uses estimation of factors including the number of stars with planets, and the lifetime of extraterrestrial civilizations.

dwarf star: Stars on the main sequence of the HR diagram.

electromagnetic spectrum: The full range of wavelengths of electromagnetic radiation. Long (low energy: think radio and microwaves) wavelengths are at one end of the spectrum, with short (high energy, such as x-rays and gamma rays) wavelengths at the other.

electron: A type of subatomic particle present orbiting atomic nuclei; is negatively charged.

emission nebula: A type of nebula created from ionized gasses; emits light at various wavelengths.

event horizon: A hypothetical surface (sometimes a sphere) around a black hole beyond which radiation and light cannot escape due to the black hole's gravity.

exoplanet: A planet in orbit around a star outside of our solar system (the planets in our solar system all orbit our Sun).

extrasolar: Cosmic activity taking place outside of our solar system.

galaxy: A collection of cosmic dust, gas, stars, stellar remnants, dark matter, and a supermassive blackhole that are bound together via gravity.

galaxy cluster: An area in space where hundreds or thousands of galaxies are held together by gravity; consists of galaxies, dark matter, and scorching-hot plasma.

gamma-ray burst (GRB): Short flare-ups of gamma rays, the most energetic radiation at the far end of the electromagnetic spectrum; sources include the merger of neutron stars.

general relativity: The theory, published by Albert Einstein in 1915, postulating that objects distort the space-time fabric through the force of gravity.

giant star: Very bright, large stars; above the main sequence of the HR diagram.

globular cluster: Large, gravitationally bound clusters containing up to millions of stars.

Goldilocks zone: Also called the *habitable zone*, the area surrounding a star with temperatures allowing for liquid water at the surface of a planet.

gravitational collapse: The compression or collapse of a celestial object due to internal gravitation; in the life cycle of a star, one of the final stages when a star runs out of fuel and collapses internally.

gravitational lensing: The phenomenon where space-time is curved, and light is bent as it would be by a lens; most often used to study objects very far from Earth.

gravity: The force of attraction between two objects, proportional to their masses and inversely proportional to the distance between them.

Hertzsprung-Russell diagram: A two-dimensional graph that plots the luminosity (absolute magnitude) of a star versus its spectral type; also called HR diagram.

Hubble constant: The current-day expansion rate of the universe. It states how much a megaparsec of space will expand in kilometers each second. Observations using variable stars and type 1a supernovae yield a result of 73 to 74 km/s/Mpc, whereas observations using the cosmic microwave background give a value of 67 km/s/Mpc. This discrepancy is called the *Hubble tension*.

interferometry: A scientific technique of using wave interference to combine the light from multiple telescopes into a single optical system.

isotropic: A condition in which a property has equal values as measured in different directions.

Kepler's Laws of Planetary Motion: Three ways of describing how planets orbit their star (or moons orbit a planet); these include

- Planets moving in elliptical orbits where the Sun is one of the foci of that ellipse.

- A line drawn between the Sun and a planet will sweep out an equal area in an equal amount of time. Put differently, planets move faster when closer to their stars and slower when farther away in a mathematically predictable way.

- The orbital period of a planet is related mathematically to the size of its orbit as $P^2 \propto a^3$ (P = orbital period, a = semi-major axis of the ellipse).

light-year: The astronomical distance that light travels in one year; 5.9×10^{12} miles (9.467×10^{12} km).

magnetosphere: The region in space surrounding a planet that's affected by its magnetic field.

mass: The amount of matter an object contains.

momentum: The impetus an object gains as it moves; is related to velocity and mass by $p = mv$ (where p = momentum, m = mass, and v = velocity).

nebula: Clouds of interstellar dust and gas; are often star-forming regions or surround stellar remnants.

neutron: A type of subatomic particle present in most atomic nuclei; has no electric charge.

Newton's laws of motion: Sir Isaac Newton's three ways of describing the motion and interaction of forces. These include

- Objects remain as they are, at rest or in motion, unless an external force acts on them.

- The sum of the forces on an object is equal to its mass times acceleration *(F=ma)*.

- Each action has an equal and opposite reaction.

nova: A star, such as a white dwarf, that undergoes an explosion, displays a sudden increase in brightness, then fades back to its former level of brightness.

nucleosynthesis: The formation of elements through a variety of different methods including the fusion of smaller elements at the heart of a star; stars initially create energy via nucleosynthesis as hydrogen is fused into helium.

open star cluster: Groupings of stars, typically under a few thousand, that are loosely held together by gravity.

orbit: A repeating path one object takes as it follows a gravitationally bound path around at least one other object.

planetary nebula: A type of emission nebula; sphere or hourglass shape created from the expelled atmosphere shed by a dying star, which may look like a planet through a small telescope. Is not actually a planet.

planetesimal: A celestial object created from gas and cosmic dust in a protoplanetary nebula surrounding a newly forming star; possibly impacted with others to form planets as part of a solar system's evolution.

proton: A type of subatomic particle present in atomic nuclei; is positively charged.

protostar: A cloud of gas and dust that is in the early phases of star formation; eventually collapses and contracts into a star.

pulsar: A rapidly rotating neutron star; a celestial object that emits radiation as it rotates and appears to pulse as viewed from Earth.

quasar: A very large, bright active galactic nucleus; emits massive amounts of radiation energy, and is most often found in the distant/earlier universe.

radial velocity: Also called *line of sight velocity*, the velocity of a star as it moves along the observer's line of sight. The radial velocity technique can be used to detect exoplanets by observing their parent star's wobbles.

red dwarf: A low-luminosity and cooler main-sequence star.

redshift: A way of describing the change in an electromagnetic signal's wavelength as it moves farther away from the observer; as the light waves stretch farther from us, they get longer and move toward the red side of the electromagnetic spectrum.

reflection nebula: A type of nebula that reflects the light scattered from nearby stars.

Schwarzchild radius: The radius at which the escape velocity of an object is equal to the speed of light.

solar flare: High-energy bursts of radiation from the surface of the Sun; often associated with sunspots.

solar mass: The mass of the Sun, used as an astronomical unit for comparison throughout the solar system; 4.38×10^{30} pounds (1.99×10^{30} kg).

solar wind: A steady flow of electrically charged particles emitted from the Sun's corona (outermost layer of its atmosphere).

special relativity: The theory, published by Albert Einstein in 1905, proposing a fundamental relationship with speed, mass, space, and time; includes the famous formula $E = mc^2$ where E = energy, m = mass, and c = the speed of light.

standard candle: A class of objects in astrophysics that have a known luminosity.

star formation region: An area in interstellar space with dense molecular clouds that have begun to collapse and compress via gravity, leading to the birth of new stars.

supernova: A star explosion caused by changes in the star's core, such as a large star running out of fuel, or changes caused by adding too much mass to a white dwarf.

supernova remnant: A kind of nebula formed from the remaining material when a supernova explodes and the shock wave of the supernova interacts with surrounding material.

terrestrial planet: Planets that are composed of materials similar to Earth, including metal and rock, and have a hard or oceanic surface.

transit: In astronomy, the act of one object crossing or passing in front of another and blocking light.

transit timing variation (TTV): A way of detecting exoplanets by looking at variations, or changes, in the timing of a planet's transit in front of a star.

vacuum: Any location where the pressure is low enough that molecules are too far spread out to interact.

variable star: A star whose apparent brightness changes, over either long or short periods of time.

white dwarf: The dense core of a star that initially had eight solar mass or less that's run out of nuclear fuel and has shed its exterior layers into a planetary nebula.

Wien's law: Also called Wien's displacement law; frames a relationship between black body temperature and the wavelength of its emitted light.

WIMPS: Weakly interacting massive particles, hypothetical particles that may help explain dark matter.

Index

N

naming
  astronomical objects, 62–63
  exoplanets, 134
Nancy Grace Roman Space Telescope, 140,
    141, 352
NASA
  Europa Clipper, 149–150
  Swift Observatory, 238
NASA Exoplanet Catalog (website), 129
nebula
  about, 125–126
  dark, 126–127
  defined, 356
  diffuse, 126–127
  planetary, 127–128
Neptune, 56, 130, 133, 143
Neptune-like exoplanets, 133, 134–135
neutral hydrogen
  formation of atoms, 275–276
  hydrogen compared with, 276
neutrinos, 78, 220
neutron stars, 164–166
neutrons, 22–24, 356
Newton, Isaac (scientist), 38–39, 159–160,
    221, 290–291, 308
Newtonian gravitational constant, 294
Newtonian mechanics, 292
NGC 185 galaxy, 213
NGC 1566 galaxy, 200
NGC 2865 galaxy, 185
NGC 5813 galaxy, 217
NGC 6861 galaxy, 186
Norma arm, 176
normal matter, 191
nova, 105–106, 356

nuclear energy, 35
Nuclear Epoch, 265
nuclear forces, 34
nuclear fusion, 278
nucleosynthesis, 102, 106–107, 356

O

observational astronomy, 59–63
observational astrophysics, 72
observational techniques/biases, 137
observatories, 16, 76. *See also specific
    observatories*
Omega value (Ω), of universe, 316–323
Omicron Ceti, 103
Oort Cloud, 130
open star clusters, 121–122, 356
open systems, 218
optical observatories, 76
optical telescopes, 74–76
orbital period, 70–71, 112
orbits
  about, 70–71
  defined, 356
  science of, 114
Orion (constellation), 62, 158
Orion Spur, 177
outer gas giant planets, 130
outer solar system, 54
oxygen, 281

P

parallel universes, 298–299
Parker Solar Probe, 80, 82–83
partial lunar eclipse, 84
particle physics, 235
particles

Omega value of, 316–323
parallel, 298–299
ripples in, 276
static, 297
Uranus, 56, 130, 143, 221
UY Scuti (star), 154

V

vacuum, 27, 81, 358
vacuum decay, 334–335
variable stars, 103–104, 358
Vaucouleurs, Gérard de (astronomer), 175
velocity, 40
Venus, 18, 56, 130, 133
Vera Rubin Observatory, 301
Very Large Array (VLA), 77
Very Large Telescope, 141
Virgo detector, 164
Virgo Supercluster, 198, 218
virial theorem, 210–211
virtual particles, 308
visible light (optical), 12
visible wavelength observations, 74
visual binary, 116
volume, 27, 28

W

Warning icon, 4
WASP-18b, 146
WASP-39b, 146
water, on planets, 132
water molecule, 24, 25, 28–29
wavelengths, 13–14

waves, 296
weak anthropic principle, 332
weak force, 31, 34
weak lensing, 244–245
weakly interacting massive particles
(WIMPS), 219–220, 302, 358
weather, in space, 81
Webb, James (government official), 346
websites
Cheat Sheet, 4
International Astronomical Union
(IAU), 134
NASA Exoplanet Catalog, 129
weight, 28
white dwarf, 102, 105, 128, 154–155, 358
Wien's law, 358
Wilkinson Microwave Anisotropy Probe
(WMAP), 258, 320–321
Wilson, Robert (astronomer), 257–258
work, 39
wormholes, 245–247

X

XMM-Newton X-ray Observatory, 203, 228
X-ray binary star systems, 115
x-ray burster, 115–116
x-ray spectroscopy, 228
x-rays, 12, 216–218

Z

Zel'dovich, Yakov (astrophysicist), 212
zero eccentricity, 114
Zwicky, Fritz (astronomer), 211–212, 300

About the Authors

Cynthia Phillips, PhD, is a scientist at the NASA Jet Propulsion Laboratory in Pasadena, California She received her AB in astronomy, astrophysics, and physics at Harvard University and her PhD in planetary science with a minor in geosciences from the University of Arizona. She worked at the SETI Institute for 15 years prior to her current role at NASA, and is presently working on a number of different projects, including the Europa Clipper mission to Jupiter's ocean moon, which will launch in 2024. Dr. Phillips lectures on astronomy and planetary science to audiences ranging from elementary school students to science teachers to the general public, including at events such as SXSW and San Diego Comic-Con. She particularly enjoys sharing the excitement of the cosmos with the general public.

Shana Priwer has an undergraduate degree from Columbia University in architecture with minors in math and art history. She earned her master's degree in architecture from Harvard University and currently works with a software company based in San Francisco. In her role as a software development engineer in test, Shana spends much of her time understanding use cases and workflows, then using that foundational knowledge to probe, find flaws, and otherwise contribute to creating the best possible customer experience. The concern for the end user's experience is what also drives Shana in her writing. Her goal for this book was to take what's arguably some very complicated, science-heavy material and make it readable, enjoyable, and understandable. We'll leave it to you, the reader, to tell us how we did!

Cynthia and Shana are the co-authors of about a dozen books, including *Space Exploration for Dummies*, *101 Things You Didn't Know About Einstein*, and *The Everything Da Vinci Book*. They are also co-authors of the Frameworks series on the science behind the structures that are part of our built environment.

Dedication

We dedicate this book to our five children: Zoecyn, Elijah, Benjamin, Sophia, and Isaac. You are the most important people in our lives, and we love you all to the end of the cosmos and back.

Authors' Acknowledgments

We offer our heartfelt thanks to our children for their patience, support, and willingness to let us focus on the creation of this book.

A big thank you is extended to our editor, Christopher Morris, for his support and general assistance along the way.

We also offer our great thanks to our technical editor, Dr. Pamela Gay, for her amazing support in making sure we got all the details just right.

Finally, we'd like to thank our acquisitions editor Lindsay Lefevere Berg for offering us the opportunity to work on this project.

Publisher's Acknowledgments

Acquisitions Editor: Lindsay Lefevere Berg

Project Editor: Christopher Morris

Copy Editor: Christopher Morris

Technical Editor: Pamela Gay

Production Editor: Saikarthick kumarasamy

Cover Image: Courtesy of NASA and the Space Telescope Science Institute (STScI) — public domain

Leverage the power

Dummies is the global leader in the reference category and one of the most trusted and highly regarded brands in the world. No longer just focused on books, customers now have access to the dummies content they need in the format they want. Together we'll craft a solution that engages your customers, stands out from the competition, and helps you meet your goals.

Advertising & Sponsorships

Connect with an engaged audience on a powerful multimedia site, and position your message alongside expert how-to content. Dummies.com is a one-stop shop for free, online information and know-how curated by a team of experts.

- Targeted ads
- Video
- Email Marketing
- Microsites
- Sweepstakes sponsorship

20 MILLION PAGE VIEWS EVERY SINGLE MONTH

15 MILLION UNIQUE VISITORS PER MONTH

43% OF ALL VISITORS ACCESS THE SITE VIA THEIR MOBILE DEVICES

700,000 NEWSLETTER SUBSCRIPTIONS TO THE INBOXES OF *300,000* UNIQUE INDIVIDUALS EVERY WEEK

# of dummies

Custom Publishing

Reach a global audience in any language by creating a solution that will differentiate you from competitors, amplify your message, and encourage customers to make a buying decision.

- Apps
- Books
- eBooks
- Video
- Audio
- Webinars

Brand Licensing & Content

Leverage the strength of the world's most popular reference brand to reach new audiences and channels of distribution.

For more information, visit **dummies.com/biz**

PERSONAL ENRICHMENT

 **Staying Sharp**
9781119187790
USA $26.00
CAN $31.99
UK £19.99

 **Facebook**
9781119179030
USA $21.99
CAN $25.99
UK £16.99

 **Guitar**
9781119293354
USA $24.99
CAN $29.99
UK £17.99

 **Investing**
9781119293347
USA $22.99
CAN $27.99
UK £16.99

 **Beekeeping**
9781119310068
USA $22.99
CAN $27.99
UK £16.99

 **Digital Photography**
9781119235606
USA $24.99
CAN $29.99
UK £17.99

 **Meditation**
9781119251163
USA $24.99
CAN $29.99
UK £17.99

 **Pregnancy**
9781119235491
USA $26.99
CAN $31.99
UK £19.99

 **Samsung Galaxy S7**
9781119279952
USA $24.99
CAN $29.99
UK £17.99

 **iPhone**
9781119283133
USA $24.99
CAN $29.99
UK £17.99

 **Crocheting**
9781119287117
USA $24.99
CAN $29.99
UK £16.99

 **Nutrition**
9781119130246
USA $22.99
CAN $27.99
UK £16.99

PROFESSIONAL DEVELOPMENT

 **Windows 10**
9781119311041
USA $24.99
CAN $29.99
UK £17.99

 **AutoCAD**
9781119255796
USA $39.99
CAN $47.99
UK £27.99

 **Excel 2016**
9781119293439
USA $26.99
CAN $31.99
UK £19.99

 **QuickBooks 2017**
9781119281467
USA $26.99
CAN $31.99
UK £19.99

 **macOS Sierra**
9781119280651
USA $29.99
CAN $35.99
UK £21.99

 **LinkedIn**
9781119251132
USA $24.99
CAN $29.99
UK £17.99

 Windows 10
9781119310563
USA $34.00
CAN $41.99
UK £24.99

 **SharePoint 2016**
9781119181705
USA $29.99
CAN $35.99
UK £21.99

 **Fundamental Analysis**
9781119263593
USA $26.99
CAN $31.99
UK £19.99

 **Networking**
9781119257769
USA $29.99
CAN $35.99
UK £21.99

 **Office 2016**
9781119293477
USA $26.99
CAN $31.99
UK £19.99

 Office 365
9781119265313
USA $24.99
CAN $29.99
UK £17.99

 Salesforce.com
9781119239314
USA $29.99
CAN $35.99
UK £21.99

 **Coding**
9781119293323
USA $29.99
CAN $35.99
UK £21.99

dummies.com

dummies
A Wiley Brand